Genetics and Animal Breeding

Genetics and Animal Breeding

Dr. Swarnim

AMIGA PRESS. INC
Delhi -110051

AMIGA PRESS. INC
H.O :- 28-A, Indra Park, Chander Nagar
Delhi :- 110051 (India)
Tel :- 011-22504028, +91-98736-57000
Email :- amigapress@gmail.com

This edition is published by Amiga Press.Inc

ISBN : 978-81-929542-7-1

First Published 2018

PRINTED : REPLIKA PRESS PVT. LTD.

Preface

The science of Animal breeding is defined as the application of the principles of Genetics and biometry to improve the efficiency of production in farm animals. These principles were applied to change animal populations thousands of years before the sciences of genetics and biometry were formally established. The practice of animal breeding dates back to the Neolithic period, when people attempted to domesticate wild species such as reindeer, goats, hogs and dogs. Domestication was performed through controlled mating and reproduction of captive animals which were selected and mated based on their behaviour and temperament. Judging from cave paintings that have survived, selection was also applied to some qualitative traits such as coat colour and the absence or presence of horns.

Mating animals of the same breed for maintaining such breed is referred to as purebred breeding. Opposite to the practice of mating animals of different breeds, purebred breeding aims to establish and maintain stable traits, that animals will pass to the next generation. By "breeding the best to the best," employing a certain degree of inbreeding, considerable culling, and selection for "superior" qualities, one could develop a bloodline or "breed" superior in certain respects to the original base stock.

Artificial insemination (AI) now accounts for around 75 per cent of British cattle births. The cow is tied up and one hand of the inseminator manipulates the cervix through the rectum wall whilst the other discharges semen into the vagina and cervix using an inseminating gun. A regional manager with AI firm Genus the problems associated with the procedure. "Placing a quarter of a cc of diluted semen into an area as small as a penny coin, when the target area is inside 500kg of unco-operative cow, requires a high degree of expertise. The more the cow is able to move about, the harder it becomes to hit the target.

Poor handling facilities cause stress, the cow, the inseminator and both our bank balances." There are many invasive procedures carried out on farm animals that the general public rarely witness. One of many includes making sure the cow is in calf. This is carried out using an ultrasound device much like a small TV set with a long thick wire attachment. A vet will insert their hand and arms length into the vagina of the cow.

The animal is held in a holding device to keep her still. The screen shows the small calf in the middle of a dark shaped uterus. Herds usually calve at the same time (due to human intervention) so all the cows on the farm will usually be checked at the same time.

Modern animal breeders use a lot of information to select which animals will become parents of the next generation. They consider the records they have kept for their animals. They might want to increase dairy cow milk production or wool production in sheep. The goal of their breeding program then becomes to produce animals that have these good traits.

The basic concepts of this book imparts rudimentary knowledge for the students and research scholars.

—*Editor*

Contents

1

Animal Breeding Methods

The science of Animal breeding is defined as the application of the principles of Genetics and biometry to improve the efficiency of production in farm animals. These principles were applied to change animal populations thousands of years before the sciences of genetics and biometry were formally established.

The practice of animal breeding dates back to the Neolithic period, when people attempted to domesticate wild species such as reindeer, goats, hogs and dogs. Domestication was performed through controlled mating and reproduction of captive animals which were selected and mated based on their behaviour and temperament. Judging from cave paintings that have survived, selection was also applied to some qualitative traits such as coat colour and the absence or presence of horns.

Without written records, there is no certain knowledge of the evolution of animal breeding practices, but written documents dating back more than 4000 years indicate that humans appreciated the significance of family resemblance in mating systems, recognized the dangers of intense inbreeding, and used castration to prevent the reproduction of undesirable males. Progress in the performance of domesticated animals through these selection practices was very slow; improvements were mainly due to animals adapting better to their environments.

Robert Bakewell, an English animal breeder of the 18th century, is considered the founder of systematized animal breeding. He was the first to emphasize the importance of accurate breeding records, introduced the concept of progeny testing to evaluate the genetic potentials of young sires, and applied inbreeding to stabilize desired qualitative traits. He also promoted concepts such as "like begets like," "prepotency is associated with inbreeding" and "breed the best to the best." Bakewell and his contemporaries in Europe pioneered the development of diverse breeds of Beef cattle, Dairy cattle, Sheep. Colour, conformation, geographical origin and some production characteristics were the main factors that differentiated these breeds. Wide geographical redistribution of animal populations was also an important factor in the formation of new breeds, as invading armies, migrating people and traders transported livestock to new lands.

Animal breeding as a modern science belongs to the 20th century. Although numerous geneticists and biometricians have made significant contributions to the development of this science, J.L. Lush of Iowa State University is considered the father of the modern science of animal breeding. Lush and his students developed major scientific procedures applicable to the genetic improvement of farm animals.

COW BREEDING

Different Breeds for Different Needs

Dairy cattle are breeds that have been developed to produce milk whereas beef cattle are bred and selected primarily for the production of meat. Dual-purpose breeds do exist and they have been selected for both meat and milk production. In the last 20 years the genetic make up of the UK dairy herd has changed and is still changing. Cows are larger, earlier maturing, are producing far more milk and are more efficient at converting feed into milk.

As it is the birth of the calf that causes milk production, it is critical for the farmer that the cow becomes pregnant again during lactation and produces the next calf one year after the previous one. The average length of productive life of a dairy cow is between four and five lactations, although this varies from one or two to over ten. Since the fifties artificial insemination (AI) has increasingly replaced natural service. Embryo transplants (ET) are also being developed. A surrogate mother may have difficulty delivering, since the foreign calf is often larger than nature intended. When this happens, a rope or chain is attached to the calf's forelegs, and, with a winch-like device the newborn calf is pulled from the womb. Unnecessary suffering for both cow and calf.

Regular Calving

A high annual milk production per cow is dependent on regular calving at about 12- month intervals. Poor reproductive efficiency leads to longer calving indices and a reduction in milk sales per cow per annum. It also leads to an increase in the number of cows culled for failing to conceive. On average, in the UK, only about 85 cows calve per annum for every 100 cows available for breeding. Over thirty-five per cent of all cows culled are disposed of because of reproductive disorders.

Obtaining Semen

Photographer Henry Heap visited the Milk Marketing Board stud bull operation in Mid Wales. He provided Turning Point magazine with a complete set of photographs and commentary of his visit. "A large, handsome, tan bull stood stock still, tethered by the ring tethered through his nose to what looked like a giant circular clothes line or some grotesque merry-go-round. He was kept so motionless by a rubber hood that covered his eyes, which had a similar effect to that of hooding a falcon.

A worker commented that the bull would stay still all day if they left the blindfold on. However, this bull was not the source of the semen, but instead what is known as a 'teaser'. The stud bulls are led out by the nose one at a time to mount the motionless hooded bull. After they have mounted the teaser their penis is diverted into a leather vagina which is used to collect the semen. As soon as the process is complete, another bull is led out. The hood keeps the teaser bull stationery for each successive beast to mount. These stud bulls never even see females.

During embryo transfer an embryo can be flushed from the uterus of a genetically superior cow and implanted in a receiver cow of so-called 'poorer quality' for foetal development. Embryos are also taken from dead cows at the slaughterhouse. Hormone treatment is used to induce multiple ovulations and the cow is then inseminated. A few days later several embryos can be flushed from her.

For both embryo collection and implanting the cow receives a local anaesthetic injection in the tail. Embryos of the desired breed can then be used. ET maximises the number of offspring from selected breeding females and maximises profit. Do-it-yourself embryo transfer is now taking place in the UK. Courses are run to train in direct embryo transfer techniques using the quick-thaw method for frozen embryos. Although vets are not required to be present at the time of transfer, they must check each animal is suitable for implantation. Farmers genetically improve the herd by using techniques such as MOET (multiple ovulation embryo transfer) which involves inducing a genetically 'superior' cow to produce up to 20 embryos rather than one.

For ease of management, farmers prefer cows to calf at the same time. They, therefore, need to detect when cows come into oestrous. The oestrous detection rate (ODR) is the number of cows observed in oestrus during a 3 week period, in relation to the number of cows expected to be observed. The average ODR on farms is about 60 per cent. One alternative is to synchronise the oestrous period of several animals by injections of prostaglandin or administration of progesterone as an impregnated intraveginal coil implanted for 9-12 days.

Artificial Insemination

Artificial insemination (AI) now accounts for around 75 per cent of British cattle births. The cow is tied up and one hand of the inseminator manipulates the cervix through the rectum wall whilst the other discharges semen into the vagina and cervix using an inseminating gun. A regional manager with AI firm Genus the problems associated with the procedure. "Placing a quarter of a cc of diluted semen into an area as small as a penny coin, when the target area is inside 500kg of unco-operative cow, requires a high degree of expertise. The more the cow is able to move about, the harder it becomes to hit the target.

Poor handling facilities cause stress, the cow, the inseminator and both our bank balances." There are many invasive procedures carried out on farm animals that the general public rarely witness. One of many includes making sure the cow is in calf. This is carried out using an ultrasound device much like a small TV set with a long thick wire attachment. A vet will insert their hand and arms length into the vagina of the cow.

The animal is held in a holding device to keep her still. The screen shows the small calf in the middle of a dark shaped uterus. Herds usually calve at the same time (due to human intervention) so all the cows on the farm will usually be checked at the same time.

Jersey Cattle

A small, honey-brown breed of dairy cattle, the Jersey is renowned for the high butterfat content of its milk, as well as a genial disposition. The Jersey cow is quite small, ranging from only 800 to 1200 pounds (360 to 540 kg).

The main factor contributing to the popularity of the breed has been their greater economy of production, due to:

- The ability to carry a larger number of effective milking cows per unit area due to lower body weight, hence lower maintenance requirements
- High butterfat conditions, and to thrive on locally produced food.

Bulls are also small, ranging from 1200 to 1800 pounds (540 to 820 kg), and are notoriously very aggressive. Castrated males can be trained into fine oxen which, due to their small size and gentle nature make them popular with young teamsters. Jersey oxen are not as strong as larger breeds however and are generally out of favour among competitive teamsters. Due to the small size, curious docile character and attractive features of the Jersey cow, small herds were imported into England by aristocratic landowners as props for aesthetic landscapes. As its name implies, the Jersey was bred on the British Channel Island of Jersey.

It apparently descended from cattle stock brought over from the nearby Norman mainland, and was first recorded as a separate breed around 1700.

Frank Falle has speculated (on the basis of DNA evidence) that the oldest settlers in Jersey were Danish vikings who had been to Nantes with Hatain, where it is recorded that tribute was given by the King of France to Hatain of 500 cattle to leave that area, whereupon they settled in the Normandy area. The marked resemblance between Jersey cattle and the Nantaise cow would seem to bear this out. While the breed is isolated from outside influence today, this was not always the case. Before 1789 cows would be given as dowry for inter-island marriages between Jersey and Guernsey. This was, however, not widespread. Since 1789, imports of foreign cattle into Jersey have been forbidden by law to maintain the purity of the breed, although exports of cattle and semen have been an important economic resource for the island.

The restriction on the import of cattle was initially introduced in 1789 to prevent a collapse in the export price. The United Kingdom levied no import duty on cattle imported from Jersey. Cattle were being shipped from France to Jersey and then being shipped onward to England to circumvent the tariff on French cattle. The increase in the supply of cattle, sometimes of inferior quality, was bringing the price down and damaging the reputation of Jersey cattle. The import ban stabilised the price and enabled a more scientifically controlled programme of breeding to be undertaken.

Sheep Breeds

The domestic sheep is a multi-purpose animal, and the more than 200 breeds now in existence were created to serve these diverse purposes. Some sources give a count of a thousand or more breeds, but concrete documentation has not been produced for such a number. Almost all sheep are classified as being best suited to furnishing a certain product: wool, meat, milk, hides or a combination thereof (*i.e.* dual-purpose breeds). Other features that are important when classifying sheep include face colour (generally white or black), the length of the tail, the presence or lack of horns, and the topography for which the breed has been developed.

This last point is especially stressed in the United Kingdom, where breeds are described as either upland (hill or mountain) or lowland breeds. A sheep may also be of a fat-tailed breed, which is a type of dual-purpose sheep common in Africa and Asia with larger deposits of fat within its tail, or the meat from such a breed. Ovine breeds are also grouped based on how well they are suited to producing a certain type of breeding stock. Generally, sheep are thought to be either "ewe breeds" or "ram breeds". Ewe breeds are those that are hardy, and have good reproductive and mothering capabilities—they are for replacing breeding ewes in standing flocks. Ram breeds are selected for rate of weight gain and carcass quality, and are mated with ewe breeds to produce meat lambs. Many breeds, especially rare or primitive ones, fall into neither category strictly. Lowland and upland breeds are also crossed in this fashion, with the hardy hill ewes crossed with larger, fast-growing lowland rams to produce a type of ewe called a mule, which can then be crossed with meat-type rams to produce prime market lambs.

Sheep breeds are categorized by the type of wool they exhibit, irrespective of whether they are favoured for meat or wool production. Fine wool breeds are those that have wool of great crimp and density, and are consequently preferred by the textile industry.

Most of these breeds were derived from Merino sheep, and Merinos continue to dominate the world sheep industry. The record for the most valuable sheep ever belongs to an Australian Merino ram which was sold for $16,000 AUD. Medium wool breeds have wool between the extremes, and are typically fast-growing meat and ram breeds with dark faces. Some major medium wool breeds, such as the Corriedale, are dual-purpose crosses of long and fine-wooled breeds created for high-production commercial flocks. Long wool breeds are the largest of sheep, with lengthy wool and a slow rate of maturation. Long wool sheep are most valued by modern agriculture for crossbreeding to improve the attributes of other sheep types. The American Columbia breed was developed through crossing Lincoln rams—a long wool breed—with fine-wooled Rambouillet ewes.

Coarse or carpet wool sheep are those with a medium to long length wool of characteristic coarseness. Breeds traditionally used for carpet wool show more variability in general traits than other types, the chief requirement being a wool that will not break down under heavy use (as would that of the finer breeds). As the demand for carpet-quality wool declines, some breeders of this type of sheep are attempting to reposition a few of the traditional breeds to alternative purposes. Others have always been primarily meat-class sheep. A last group of sheep breeds is that of fur or hair sheep, which do not grow wool at all. Hair sheep are similar to the very early type of domesticated sheep kept before wooly breeds were developed, and are primarily raised for meat and pelts. Some modern breeds of hair sheep, such as the Dorper, are the result of crosses between wooled and hair breeds. For producers uninterested in selling wool, hair sheep are highly advantageous, as those who keep hair sheep need not go to the expense of shearing. Hair sheep are also resistant to parasites and hot weather, and are consequently increasing in popularity among a subset of producers.

With the modern rise of corporate agrobusiness and the decline of localized family farms, many breeds of sheep are in danger of extinction. Preference for only those breeds with uniform characteristics and fast growth rates has pushed heritage (or heirloom) breeds to the margins of the sheep industry. Those that remain are maintained through the efforts of conservation organizations, breed registries, and individual farmers dedicated to their preservation.

SELECTING SIRES FOR A SINGLE HERD

Most of the factors involved in sire selection, but one major question remains. Will the sire which has been proven to be superior in other herds also be superior in my herd? Older breeders will remember the term "nicking," which is not too often heard anymore. If a sire did well in one herd and poorly in a second, he was said to nick well in the first, poorly in the second. Today, researchers look at the importance of nicking by studying the importance of sire by farm interactions. Such interactions arise if there are real changes in the relative performance of sires (usually measured by the milk yield of their daughters) as they are used on various farms. Should there be an appreciable number of sires which rank high in some herds or farms, and low in others, and vice versa for other sires, this would show up as a sire-by-herd interaction.

The importance of this question is apparent; if there were such interactions, a dairymen would be unable to rely on USDA sire proofs, since he could not depend on sires with superior proofs to be superior in his herd. This was an area that obviously needed research, and during the past several years considerable research has been published on the subject. A classic study by North Carolina scientists involved sire-by-region interactions. The two regions involved were North Central and Southeastern United States and the question essentially was, "Do AI sires used in these two areas rank about the same, or are there some who are outstanding in one region but poor in the other, and vice versa?" Much of the semen used by Florida dairymen is from sires bred, house and proven in the Eastern and North Central United States, so this question was of immediate concern.

Results showed that this particular type of interaction was not an important factor, and indeed was almost non-existent. The authors stated, "provided adequate unbiased progeny information is utilized, a bull may be safely selected and used for artificial insemination in one region, even though he was evaluated in the other region." Studies on the importance of sire by herd interactions have been in very close agreement, with one being conducted by Florida researchers on records of Florida DHIA Guernsey, Holstein and Jersey first-calf heifers. Everyone recognizes that even the best bull may have a certain number of poor daughters and that a dairyman who has only a few daughters of a particular sire may have been unlucky, by chance, and gotten one or more of them. The opposite can happen also with poor sires.

The problem is to determine how important the interactions are overall, realizing that a certain amount of shifting in rankings of sires can occur by chance, particularly if a sire is represented in one or more herds by only a small number of daughters. The research found that under Florida DHIA herd conditions; these interactions were of little or no practical or theoretical consequence for milk and fat yields.

There was some evidence among Guernsey's and Jerseys of an interaction for fat percentage; this has been suggested in other research studies also, but many of the scientists who have studied interactions have not looked at fat percentage. In summary, it appears that sires proven superior for milk and fat yields in other herds can be used with confidence in Florida dairy herds, without concern for possible sire by herd interactions. The possibility exists for the presence of interactions with fat percentage.

Finding Superior Females

Measurements for most of the economically important traits can be made directly on females. It frequently takes a long time to get the measurements of interest, however. Mature body size, would have to wait until the cow is 6 to 8 years old. An estimate of milk yield must wait until the first lactation is completed (at about 3 years of age), unless one accepts an estimate of the lactation based on the first several months. These points are mentioned because genetic progress is influenced by the time at which measurements are made, and their reliability. Take milk yield as an example. An estimate of the milk producing ability of a cow comes with her first milk record.

Certainly, however, one knows more about her when there are two records rather than one, and even more with three records. Generally, the more records available, the more reliable the estimate of her real ability; but it takes longer to get more records. An estimate of the producing ability of a cow was called "Most Probable Producing Ability" years ago by Dr. J.L. Lush, well known animal geneticist. This measure takes into consideration the idea of repeatability, and shows how the number of records a cow has should influence the estimate of her producing ability. If a heifer produces more than her herdmates during her first lactation, we will certainly guess that she is a better than average animal.

If the average of her first two records is above herdmates, one would be even more confident that she is truly above average, and her good performance did not just happen from some favourable accident which probably will not happen again. The calculations are simple and can be applied to any trait of interest. Although the concept is rarely mentioned today, since availablity of computers makes more sophisticated procedures feasible, MPPA serves as the basis for modern methods. State the milk record, or any other measurement, as a deviation from herdmates. If the cow has two measurements, state each as a deviation, and take the average. MPPA then can be calculated as:

$$\text{MPPA} = (\text{Average deviation}) \times [(nr)/(1 + nr - r)]$$

where n = the number of measurements and r = the repeatability. How much better two measurements or records are than one, as far as reliability, is determined by a similar equation.

$$\text{Gain in accuracy} = 1 - [(1 + nr - r) / n]$$

with n and r as in MPPA. The more records one has on a cow, the more one can accept her records at face value. If she has six records averaging 2000 pounds better than her herdmates, the best guess of her MPPA is 1714 pounds better.

Which cow should be selected as a dam for a future herd sire, a cow with two records averaging 3000 pounds above herdmates, or one with four records averaging 2400 pounds above? The two estimates of MPPA are close, being 2000 pounds for the cow with two records and 1920 pounds for the older cow. One probably would look at other characteristics of the two animals, particularly fat per cent but possibly others, before making a decision. The more sophisticated and useful application of this concept, of course, is that it is incorporated into calculations of ERPA.

GENETIC CHANGE IN MILK YIELD

The theoretical upper limits of genetic change in milk yield have been known for quite some time. Knowing the various repeatabilities, heritabilities, generation intervals and selection pressures (intensities), one can calculate what would be expected to happen. This tells nothing about what is actually going on, of course, but it does tell what one can expect under the best conditions.

Potential

If milk yield is given maximum emphasis in the selection programme, with little or no pressure being exerted on other traits, and efficient use is made of an AI sire proving programme, the maximum attainable genetic change in milk yield is slightly over 2.0 per cent per year.

Most of this, over 90 per cent, theoretically would come directly or indirectly from sire selection. In a Holstein population averaging 15,000 pounds per cow, this would amount to 300 pounds per year. Though this estimate has appeared disappointingly small to some breeders, it is more than the actual total change which has been attained in the United States.

Observed

It is only in recent years that it has been possible to obtain reliable estimates of the actual genetic change which has occurred in milk production. A major problem in the actual measurement of such progress is that environmental and genetic changes are confounded in such a manner over time as to make their separation extremely difficult. With the availability of high-speed large-capacity electronic computers, many of the problems can now be handled. Use of the animal model has improved the accuracy of the estimates. As far as progress arising from selection of males and females, theoretical estimates has been right on target.

About 90 to 92 per cent of all genetic change in milk yield has occurred from sire selection; this represents the selection of males and females to produce herd sires and the selection of males to produce herd dams. The remaining 8 to 10 per cent of genetic progress arises from selection of females to produce herd dams. The fact that we must raise nearly all female calves born alive in a population to maintain population size, and the difficulties of evaluating the true genetic merit of dams, precludes much selection pressure from selection of dams.

Average annual genetic change in milk yield of tested dairy cattle has been variable but seems to be about 0.7 per cent of their average levels. This would amount to about 105 pounds per cow in a 15,000 pound herd, or 140 pounds in a 20,000 pound herd. In individual herds, one would expect to find genetic changes which were considerably higher or lower. If 2.0 per cent annual genetic change is possible, as theory suggests, dairymen are progressing genetically only about one-third as fast as possible.

HORSE BREEDING

Horse breeding refers to reproduction in horses, and particularly the human-directed process of selective breeding of animals, particularly purebred horses of a given breed. While feral and wild horses breed successfully without human assistance, planned matings can be used to produce specifically desired characteristics in domesticated horses. Furthermore, modern breeding management and technologies can increase the rate of conception, a healthy pregnancy, and successful foaling.

The male parent of a horse, a stallion, is commonly known as the sire and the female parent, the mare, is called the dam. Both are genetically important, as each parent provides half of the genetic makeup of the ensuing offspring, called a foal. (Contrary to popular misuse, the word "colt" refers to a young male horse only; "filly" is a young female.) Though many amateur horse owners may simply breed a family mare to a local stallion in order to produce a companion animal, most professional breeders use selective breeding to produce individuals of a given phenotype, or breed. Alternatively, a breeder could, using individuals of differing phenotypes, create a new breed with specific characteristics.

Some breeders consider the quality of the sire to be more important than the quality of the dam. However, other breeders maintain that the mare is the most important parent. Because stallions can produce far more offspring than mares, a single stallion can have a greater overall impact on a breed. However, the mare may have a greater influence on an individual foal because its physical characteristics influence the developing foal in the womb and the foal also learns habits from its dam when young.

Foals may also learn the "language of intimidation and submission" from their dam, and this imprinting may affect the foal's status and rank within the herd. Many times, a mature horse will achieve status in a herd similar to that of its dam; the offspring of dominant mares become dominant themselves.

Estrus Cycle of the Mare

The estrous cycle controls when a mare is sexually receptive towards a stallion, and helps to physically prepare the mare for conception. It generally occurs during the spring and summer months, although some mares may be sexually receptive into the late fall, and is controlled by the photoperiod (length of the day), the cycle first triggered when the days begin to lengthen.

The estrus cycle lasts about 19-22 days, with the average being 21 days. As the days shorten, the mare returns to a period when she is not sexually receptive, known as anestrus. Anestrus prevents the mare from conceiving in the winter months, as that would result in her foaling during the harshest part of the year, a time when it would be most difficult for the foal to survive.

This cycle contains 2 phases:

1. *Estrus, or Follicular, phase*: 5-7 days in length, when the mare is sexually recepetive to a stallion. Estrogen is secreted by the follicle. Ovulation occurs in the final 24-48 hours of estrus.
2. *Diestrus, or Luteal, phase*: 14-15 days in length, the mare is not sexually recepetive to the stallion. The corpus luteum secretes progesterone.

Although 16 per cent of mares have double ovulations (allowing them to twin), this does not affect the length of time of estrus or diestrus.

Effects on the Reproductive System During the Estrous Cycle

Changes in hormone levels can have great affects on the physical characteristics of the reproductive organs of the mare, thereby preparing, or preventing, her from conceiving.

- *Uterus*: Increased levels of estrogen cause edema within the uterus, making it feel heavier, and the uterus loses its tone. This edema decreases following ovulation, and the muscular tone increases. High levels of progesterone do not cause edema within the uterus. The uterus becomes flaccid during anestrus.

- *Cervix*: The cervix relaxes right before estrus occurs, and the secretions of the cervix increase. High progesterone levels (during diestrus) cause the cervix to close.
- *Vagina*: The portion of the vagina near the cervix becomes engorged with blood right before estrus. The vagina becomes relaxed and secretions increase.
- *Vulva*: Relaxes right before estrus begins. Becomes dry, and closes tightly, during diestrus.

Hormones Involved in the Estrus Cycle, During Foaling, and after Birth

The cycle is controlled by several hormones which regulate the estrous cycle, the mare's behaviour, and the reproductive system of the mare. The cycle begins when the increased day length causes the pineal gland to reduce the levels of melatonin, thereby allowing the hypothalamus to secrete GnRH.

- *GnRH (Gonadotropin releasing hormone):* Secreted by the hypothalamus, causes the pituitary to release of 2 gonadotrophins: LH and FSH.
- *LH (Luteinizing hormone):* Levels are highest 2 days following ovulation, then slowly decrease over 4-5 days, dipping to their lowest levels 5-16 days after ovulation. Stimulates the follicle to develop, which in turn causes an increase in estrogen. Unlike most mammals, the mare does not have an increase of LH right before ovulation.
- *FSH (Follicle-stimulating hormone):* Secreted by the pituitary, causes the ovarian follicle to develop. Levels of FSH rise slightly at the end of estrus, but have their highest peak about 10 days before the next ovulation. FSH is inhibited by inhibin, at the same time LH and estrogen levels rise, which prevents immature follicles from continuing their growth.
- *Estrogen:* Secreted by the developing follicle, it causes the pituitary gland to secrete more LH (therefore, these 2 hormones are in a positive feedback loop). Additionally, it causes behavioural changes in the mare, making her more recpetive towards the stallion, and causes physical changes in the cervix, uterus, and vagina to prepare the mare for conception. Estrogen peaks 1-2 days before ovulation, and decreases within 2 days following ovulation.
- *Inhibin:* Secreted by the developed follicle right before ovulation, "turns off" FSH, which is no longer needed now that the follicle is larger.
- Progesterone: Prevents conception and decreases sexual receptibility of the mare to the stallion. Progesterone is therefore lowest during the

estrus phase, and increases during diestrus. It decreases about 15 days after ovulation, when the corpus luteum begns to decrease in size.

- Prostaglandin: Secreted by the endrometrium 13-15 days following ovulation, causes luteolysis and prevents the corpus luteum from secreting progesterone
- eCG (equine chorionic gonadotropin - formerly called PMS/PMSG pregnant mare serum and pregnant mare serum gonadotropin): Chorionic gonadotropins secreted if the mare conceives. First secreted by the endometrial cups around the 36th day of gestation, peaking around day 60, and decreasing after about 120 days of gestation. Also help to stimulate the growth of the fetal gonads.
- *Prolactin*: Stimulates lactation
- *Oxytocin*: Stimulates the uterus to contract

While horses in the wild mate and foal in mid to late spring, many horses domestically bred for competitive purposes, especially horse racing and various futurities, are usually born as close to January first as possible, so as to be at an advantage in size and maturity when competing against other horses in the same age group.

When an early foal is desired, barn managers will put the mare "under lights" by keeping the barn lights on in the winter to simulate a longer day, thus bringing the mare into estrus sooner than she would in nature. Mares signal estrus and ovulation by urination in the presence of a stallion, raising the tail and revealing the vulva. A stallion, approaching with a high head, will usually nicker, nip and nudge the mare, as well as sniff her urine to determine her readiness for mating.

Once fertilized, the oocyte (egg) remains in the oviduct for approximately 5.5 more days, and then descends into the uterus. The initial single cell combination is already dividing and by the time of entry into the uterus, the egg might have already reached the blastocyst stage. The gestation period lasts for about eleven months, or about 342 days (normal average range 320-370 days). During the early days of pregnancy, the embryo (up to about 35 days), or fetus (after 35-40 days), is mobile, moving about in the uterus until about day 16 when "fixation" occurs.

True implantation - invasion into the endometrium of any sort - does not occur until about day 35 of pregnancy with the formation of the endometrial cups, and true placentation (formation of the placenta) is not initiated until about day 40-45 and not completed until about 140 days of pregnancy. A heartbeat can be detected on day 21, and the fetus gender can be determined by day 70 of the gestation. Halfway through gestation the fetus is the size of between a rabbit and a beagle. The most dramatic fetus development occurs in the last 3 months of pregnancy when 60 per cent of fetal growth occurs.

Care of the Pregnant Mare

Domestic mares receive specific care and nutrition to ensure that they and their foals are healthy. Mares are given vaccinations against diseases such as the Rhinopneumonitis (EHV-1) virus (which can cause abortions) as well as vaccines for other conditions that may occur in a given region of the world. Mares can be used for riding or driving during most of their pregnancy, and it's healthy for them to have exercise.

But only moderate exercise, especially when they become heavy in foal. Exercise in excessively high temperatures has been suggested as being detrimental to pregnancy maintenance during the embryonic period - it should however be noted that ambient temperatures encountered during the research were in the region of 100 degrees F and the same results may not be encountered in regions with lower ambient temperatures.

Mares are generally fed more than other horses and are provided supplemental vitamins and minerals because their bodies require extra nutrition to form (and later to nurse) a foal. Many feeds designed for pregnant and lactating mares also have extra fat and protein. Overfeeding the pregnant mare should also be avoided, as excess weight may contribute to difficulties foaling or fetal/foal related problems. A special foaling stall or shed that is large and clutter free provides the mare with a safe place to give birth. Most mares foal at night or early in the morning, and prefer to give birth alone when possible. Labour is rapid, often no more than 30 minutes, and from the time the feet of the foal appear to full delivery is often only about 15 to 20 minutes. Once the foal is born, the mare will lick the newborn foal to clean it and help blood circulation. In a very short time, the foal will attempt to stand and get milk from its mother. A foal should stand and nurse within the first hour of life.

To create a bond with her foal, the mare licks and nuzzles the foal, enabling her to distinguish hers from others. Some mares are aggressive when protecting their foals, and may attack other horses or unfamiliar humans that come near their newborns. After birth, a foal's navel is dipped in iodine to prevent infection, it is sometimes given an enema to help clear the meconium from its digestive tract, and the newborn is monitored to ensure that it stands and nurses without difficulty. While most horse births happen without complications, many owners have first aid supplies prepared and a veterinarian on call in case of a birthing emergency.

People who supervise foaling should also watch the mare to be sure that she passes the placenta in a timely fashion, and that it is complete with no fragments remaining in the uterus, where they could cause a serious infection. If the placenta is not removed from the stall after it is passed, a mare will often eat it, an instinct from the wild, where blood would attract predators. Foals develop rapidly, and within a few hours a wild foal can travel with the herd. In domestic breeding, the foal and dam are usually kept separate for a while, but within a few weeks are typically pastured with the other horses. A foal will begin to eat hay, grass and grain alongside the mare at about 4 weeks old; by 10-12 weeks the foal requires more nutrition than the mare's milk can supply. Foals are typically weaned at 4-8 months of age, although in the wild a foal may nurse for a year.

How Breeds Develop

Beyond the appearance and conformation of a specific type of horse, breeders aspire to improve physical performance abilities. This concept, known as matching "form to function," has led to the development of not only different breeds, but also families or bloodlines within breeds that are specialists for excelling at specific tasks. For example, the Arabian horse of the desert naturally developed speed and endurance to travel long distances and survive in a harsh environment, and domestication by humans added a trainable disposition to the animal's natural abilities. In the meantime, in northern Europe, the locally adapted heavy horse with a thick, warm coat was domesticated and put to work as a farm animal that could pull a plow or wagon.

This animal was later adapted through selective breeding to create a strong but ridable animal suitable for the heavily-armored knight in warfare. Then, centuries later, when people in Europe wanted faster horses than could be produced from local horses through simple selective breeding, they imported Arabians and other oriental horses to breed as an outcross to the heavier, local animals. This led to the development of breeds such as the Thoroughbred, a horse taller than the Arabian and faster over the distances of a few miles required of a European race horse or light cavalry horse. Another cross between oriental and European horses produced the Andalusian, a horse developed in Spain that was powerfully built, but extremely nimble and capable of the quick bursts of speed over short distances necessary for certain types of combat as well as for tasks such as bullfighting.

Later, the people who settled the Americas needed a hardy horse that was capable of working with cattle. Thus, Arabians and Thoroughbreds were crossed on Spanish horses, both domesticated animals descended from those brought over by the Conquistadors, and feral horses such as the Mustangs, descended from the Spanish horse, but adapted by natural selection to the ecology and climate of the west. These crosses ultimately produced new breeds such as the American quarter horse and the Criollo of Argentina. In modern times, these breeds themselves have since been selectively bred to further specialize at certain tasks. One example of this is the American quarter horse. Once a general-purpose working ranch horse, different bloodlines now specialize in different events. For example, larger, heavier animals with a very steady attitude are bred to give competitors an advantage in events such as team roping, where a horse has to start and stop quickly, but also must calmly hold a full-grown steer at the end of a rope. On the other hand, for an event known as cutting, where the horse must separate a cow from a herd and prevent it from rejoining the group, the best horses are smaller, quick, alert, athletic and highly trainable. They must learn quickly, have conformation that allows quick stops and fast, low turns, and the best competitors have a certain amount of independent mental ability to anticipate and counter the movement of a cow, popularly known as "cow sense." Another example is the Thoroughbred. While most representatives of this breed are bred for horse racing, there are also specialized bloodlines suitable as show hunters or show jumpers. The hunter must have a tall, smooth build that allows it to trot and canter smoothly and efficiently. Instead of speed, value is placed on appearance and upon giving the equestrian a comfortable ride, with natural jumping ability that shows bascule and good form. A show jumper, however, is bred less for overall form and more for power over tall fences, along with speed, scope, and agility. This favours a horse with a good galloping stride, powerful hindquarters that can change speed or direction easily, plus a good shoulder angle and length of neck. A jumper has a more powerful build than either the hunter or the racehorse. The history of horse breeding goes back millennia. Though the precise date is in dispute, humans could have domesticated the horse as far back as approximately 4500 BCE. However, evidence of planned breeding has a more blurry history.

One of the earliest people known to document the breedings of their horses were the Bedouin of the Middle East, the breeders of the Arabian horse. While it is difficult to determine how far back the Bedouin passed on pedigree information via an oral tradition, there were written pedigrees of Arabian horses by A.D. 1330. The Akhal-Teke of West-Central Asia is another breed with roots in ancient times that was also bred specifically for war and racing. The nomads of the Mongolian steppes bred horses for several thousand years as well. The types of horses bred varied with culture and with the times. The uses to which a horse was put also determined its qualities, including smooth amblers for riding, fast horses for carrying messengers, heavy horses for plowing and pulling heavy wagons, ponies for hauling cars of ore from mines, packhorses, carriage horses and many others.

Medieval Europe bred large horses specifically for war, called destriers. These horses were the ancestors of the great heavy horses of today, and their size was preferred not simply because of the weight of the armor, but also because a large horse provided more power for the knight's lance. Weighing almost twice as much as a normal riding horse, the destrier was a powerful weapon in battle. On the other hand, during this same time, lighter horses were bred in northern Africa and the Middle East by Muslim warriors, who preferred a faster, more agile horse. The lighter horse suited the raids and battles of the Bedouins, allowing them to outmaneuver rather than overpower the enemy. When Muslim warriors and European knights collided in warfare, the heavy knights were frequently outmaneuvered. The Europeans, however, soon made up for the lack of speed of their native breeds by incorporating genetic traits from captured oriental horses such as the Arabian, Barb to their stables. This cross-breeding led both to a nimbler war horse, such as today's Percheron, but also to created a type of horse known as a Courser, a predecessor to the Thoroughbred, which was used as a message horse.

During the Renaissance, horses were bred not only for war, but for haute ecole riding, derived from the most athletic movements required of a war horse, and popular among the elite nobility of the time. Breeds such as the Lipizzan were developed from Spanish-bred horses for this purpose, and also became the preferred mounts of cavalry officers, who were derived mostly from the ranks of the nobility. It was during this time that gunpowder was developed, and so the light cavalry horse, a faster and quicker war horse, was bred for a "shoot and run" tactic rather than the close hand-to-hand fighting seen in the Middle Ages. After Charles II retook the British throne in 1660, horse racing, which had been banned by Cromwell, was revived. The Thoroughbred was developed 40 years later, bred to be the ultimate racehorse, through the lines of 3 foundation Arabian stallions.

The 17th and 18th centuries saw more of a need for fine carriage horses in Europe, bringing in the dawn of the warmblood. The warmblood breeds have been exceptionally good at adapting to changing times, and from their carriage horse beginnings they easily transitioned during the 1900s into a sport horse type. Today's warmblood breeds, although still used for competitive driving, are more often seen competing in the show jumping or dressage arenas. The Thoroughbred continues to dominate the horseracing world, although its lines have been more recently used to improve warmblood breeds and to develop sport horses. The predecessor of the American Quarter Horse was developed in the 1700s, mainly for quarter racing (racing ¼ of a mile). The breed was later adapted for work in the west, and "cow sense" was particularly bred for as their use for herding cattle increased.

However, because there was also a need for animals suitable for sprint racing, the modern Quarter Horse has two distinct types: the sleeker racing type and the stock horse type. The racing type most resembles the finer-boned ancestors of the first racing Quarter Horses, and the type is still used for ¼-

mile races. The stock horse type, used in western events, is bred for a shorter stride, docile temperament, and cow sense.

The need for horses for heavy draft and carriage work continued until the industrial revolution and the advent of the automobile and the tractor. After this time, draft and carriage horse numbers dropped significantly, though light riding horses remained popular for recreational pursuits. Draft horses today are used on a few small farms, but today are seen mainly for pulling and plowing competitions rather than farm work.

Heavy harness horses are now used as an outcross with lighter breeds, such as the Thoroughbred, to produce the modern warmblood breeds popular in Olympic and sport horse disciplines. Breeding a horse is an endeavor where the owner, particularly of the mare, will usually need to invest considerable time and money.

For this reason, a horse owner needs to consider several factors, including:

- Does the proposed breeding animal have valuable genetic qualities to pass on?
- Is the proposed breeding animal in good physical health, fertile, and able to withstand the vigours of reproduction?
- For what purpose will the foal be used?
- Is there a market for the foal in the event that the owner does not wish to keep the foal for its entire life?
- What is the anticipated economic benefit, if any, to the owner of the ensuing foal?
- What is the anticipated economic benefit, if any, to the owner(s) of the sire and dam or the foal?
- Does the owner of the mare have the expertise to properly manage the mare through gestation and parturition?
- Does the owner of the potential foal have the expertise to properly manage and train a young animal once it is born?

There are value judgements involved in considering whether an animal is suitable breeding stock, hotly debated by breeders. Additional personal beliefs may come into play when considering a suitable level of care for the mare and ensuing foal, the potential market or use for the foal, and other tangible and intangible benefits to the owner. If the breeding endeavor is intended to make a profit, there are additional market factors to consider, which may vary considerably from year to year, from breed to breed, and by region of the world. In many cases, the low end of the market is saturated with horses, and the law of supply and demand thus allows little or no profit to be made from breeding unregistered animals or animals of poor quality, even if registered.

The minimum cost of breeding for a mare owner includes the stud fee, and the cost of proper nutrition, management and veterinary care of the mare throughout gestation, parturition, and care of both mare and foal up to the time of weaning. Veterinary expenses may be higher if specialized reproductive technologies are used or health complications occur. Making a profit in horse breeding is often difficult. While some owners of only a few horses may keep a foal for purely personal enjoyment, many individuals breed horses in hopes of making some money in the process.

A general rule of thumb is that a foal intended for sale should be worth three times the cost of the stud fee if it were sold at the moment of birth. From birth forward, the costs of care and training are added to the value of the foal, with a sale price going up accordingly. If the foal wins awards in some form of competition, that may also enhance the price.On the other hand, without careful thought, foals bred without a potential market for them may wind up being sold at a loss, and in a worst-case scenario, sold for "salvage" value — a euphemism for sale to slaughter as horsemeat.

Therefore, a mare owner must consider their reasons for breeding, asking hard questions of themselves as to whether their motivations are based on either emotion or profit and how realistic those motivations may be.The stallion should be chosen to complement the mare, with the goal of producing a foal that has the best qualities of both animals, yet avoids having the weaker qualities of either parent. Generally, the stallion should have proven himself in the discipline or sport the mare owner wishes for the "career" of the ensuing foal. Mares should also have a competition record showing that they also have suitable traits, though this does not happen as often. A purebred horse is usually worth more than a horse of mixed breeding, though this matters more in some disciplines than others.

The breed of the horse is sometimes secondary when breeding for a sport horse, but some disciplines may prefer a certain breed or a specific phenotype of horse. Sometimes, purebred bloodlines are an absolute requirement: For example most Racehorses in the United States must be recorded with a breed registry in order to race. Bloodlines are often considered, as some bloodlines are known to cross well with others. If the parents have not yet proven themselves by competition or by producing quality offspring, the bloodlines of the horse are often a good indicator of quality and possible strengths and weaknesses.

Some bloodlines are known not only for their athletic ability, but could also carry a conformational or genetic defect, poor temperament, or for a medical problem. Some bloodlines are also fashionable or otherwise marketable, which is an important consideration should the mare owner wish to sell the foal. Horse breeders also consider conformation, size and temperament. All of these traits are heritable, and will determine if the foal will be a success in its chosen discipline. The offspring, or "get," of a stallion are often excellent indicators of his ability to pass on his characteristics, and the particular traits he actually passes on. Some stallions are fantastic performers but never produce offspring of comparable quality. Others sire fillies of great abilities but not colts. At times, a horse of mediocre ability sires foals of outstanding quality. Mare owners also look into the question of if the stallion is fertile and has successfully "settled" (*i.e.* impregnated) mares. A stallion may not be able to breed naturally, or old age may decrease his performance. Mare care boarding fees and semen collection fees can be a major cost. Breeding a horse can be an expensive endeavor, whether breeding a backyard competition horse or the next Olympic medalist.

Costs may include:

- The stud and booking fee
- Fees for collecting, handling, and transporting semen (if AI is used and semen is shipped)
- *Mare exams*: to determine if she is healthy enough to breed, to determine when she ovulates, and (if AI is used) to inseminate her
- Mare transport, care, and board if the mare is bred live cover at the stallion's residence
- Veterinary bills to keep the pregnant mare healthy while in foal
- Possible veterinary bills during pregnancy or foaling should something go wrong
- Veterinary bills for the foal for its first exam a few days following foaling

Stud fees are determined by the quality of the stallion, his performance record, the performance record of his get (offspring), as well as the sport and general market that the animal is standing for. The highest stud fees are generally for racing Thoroughbreds, which may charge from two to three thousand dollars for a breeding to a new or unproven stallion, to several hundred thousand dollars for a breeding to a stakes winner. Sport horse stallions generally range from $1000 to $3000, although the top stallions may reach $4000 for one breeding.

The lowest stud fees may only be $100-$200, but there are trade-offs: the horse will probably be unproven, and probably much less athletic than a horse with a stud fee that is a bit higher. As a stallion's career, either performance or breeding, improves, his stud fee tends to increase in proportion. If one or two offspring are especially successful, winning several stakes races or an Olympic medal, the stud fee will generally greatly increase. Younger, unproven stallions will generally have a lower stud fee earlier on in their careers.

To help decrease the risk of financial loss should the mare die or abort the foal while pregnant, many studs have a live foal guarantee (LFG), allowing the owner to have a free breeding to their stallion the next year. However, this is not offered for every breeding.

DOG BREEDING

The dog is a domesticated subspecies of the wolf, a mammal of the Canidae family of the order Carnivora. The term encompasses both feral and pet varieties and is also sometimes used to describe wild canids of other subspecies or species. The domestic dog has been one of the most widely-kept working and companion animals in human history, as well as being a food source in some cultures. There are estimated to be 400,000,000 dogs in the world. The dog has developed into hundreds of varied breeds. Height measured to the withers ranges from a few inches in the Chihuahua to a few feet in the Irish Wolfhound; colour varies from white through grays to black, and browns from light to dark in a wide variation of patterns; and, coats can be very short to many centimeters long, from coarse hair to something akin to wool, straight or curly, or smooth.

In breeding circles, a male canine is referred to as a *dog*, while a female canine is called a *bitch*. The father of a litter is called the *sire*, and the mother of a litter is called the *dam*. Offspring are generally called *pups* or *puppies* until they are about a year old. A group of offspring is a *litter*. The process of birth is *whelping*. Many terms are used for dogs that are not purebred.

Origin and Evolution

Based on DNA evidence, the wolf ancestors of modern dogs diverged from other wolves about 100,000 years ago, and dogs were domesticated from those wolf ancestors about 15,000 years ago. This date would make dogs the first species to be domesticated by humans. Evidence suggests that dogs were first domesticated in East Asia, possibly China, and some of the peoples who entered North America took dogs with them from Asia. As humans migrated around the planet a variety of dog forms migrated with them.

The agricultural revolution and subsequent urban revolution led to an increase in the dog population and a demand for specialization. These circumstances would provide the opportunity for selective breeding to create specialized working dogs and pets. This ancient mosaic, likely Roman, shows a large dog with a collar hunting a lion. Molecular systematics indicate that the domestic dog descends from one or more populations of wild wolves. As reflected in the nomenclature, dogs are descended from the wolf and are able to interbreed with wolves. The relationship between human and canine has deep roots. Converging archaeological and genetic evidence indicate a time of domestication in the late Upper Paleolithic close to the Pleistocene/Holocene boundary, between 17,000 and 14,000 years ago. Fossil bone morphologies and genetic analysis of current and ancient dog and wolf populations have not yet been able to conclusively determine whether all dogs descend from a single domestication event, or whether dogs were domesticated independently in more than one location.

Domesticated dogs may have interbred with local populations of wild wolves on several occasions (a process known in genetics as introgression). The earliest dog fossils, two crania from Russia and a mandible from Germany, date from 13,000 to 17,000 years ago. Their likely ancestor is the large northern Holarctic wolf, *Canis lupus lupus*. Remains of smaller dogs from Mesolithic cave deposits in the Middle East, dated to around 12,000 years ago, have been interpreted as descendants of a lighter Southwest Asian wolf, *Canis lupus arabs*.

Rock art and skeletal remains indicate that by 14,000 years ago, dogs were present from North Africa across Eurasia to North America. Dog burials at the Mesolithic cemetery of Svaerdborg in Denmark suggest that in ancient Europe dogs were valued companions. Genetic analyses have so far yielded divergent results. The ancestors of dogs split off from other wolves between 75,000 and 135,000 years ago, while a subsequent analysis indicated a "common origin from a single gene pool for all dog populations" between 40,000 and 15,000 years ago in East Asia. Both sets of dates must be reevaluated in light of recent findings showing that poorly calibrated molecular clocks have systematically overestimated the age of geologically recent events. On balance, and in agreement with the archaeological evidence, 15,000 years ago is the most likely time for the wolf-dog divergence.

Development of Dog Breeds

Dogs have been bred into a variety of shapes, colours and sizes. Variation can be wide even within a breed, as with these Cavalier King Charles Spaniels. There are numerous dog breeds, with over 800 being recognized by various kennel clubs worldwide. Many dogs, especially outside the United States of America and Western Europe, belong to no recognized breed. A few basic breed types have evolved gradually during the domesticated dog's relationship with humans over the last 10,000 or more years, but all modern breeds are of relatively recent derivation. Many of these are the product of a deliberate process of artificial selection. Because of this, some breeds are highly specialized, and there is extraordinary morphological diversity across different breeds. Despite these differences, dogs are able to distinguish dogs from other kinds of animal.

The definition of a dog breed is a matter of some controversy. Depending on the size of the original founding population, closed gene pool breeds can have problems with inbreeding, specifically due to the founder effect. Dog breeders are increasingly aware of the importance of population genetics and of maintaining diverse gene pools. Health testing and new DNA tests can help avoid problems, by providing a replacement for natural selection. Without selection, inbreeding and closed gene pools can increase the risk of severe health or behavioural problems. Some organizations define a breed more loosely, such that an individual may be considered of one breed as long as 75 per cent of its parentage is of that breed. These considerations affect both pets and the show dogs entered in dog shows.

Even prize-winning purebred dogs sometimes possess crippling genetic defects due to founder effect or inbreeding. These problems are not limited to purebred dogs and can affect cross-breed populations. The behaviour and appearance of a dog of a particular breed can be predicted to a degree, while mixed-breed dogs show a broader range of innovative appearance and behaviour. Mixed-breed dogs or Mongrels (also called "mutts") are dogs that do not belong to specific breeds, being mixtures more than two in variant percentages. Mixed breed dogs and purebred dogs are both suitable as companions, pets, working dogs, or competitors in dog sports.

Sometimes different breed dogs are deliberately bred, to create cross-breeds such as the Cockapoo, a mixture of Cocker Spaniel and Miniature Poodle. Such deliberate crosses may display some degree of hybrid vigour and other desirable traits, but may or may not inherit any of the desired traits of their parents, such as temperament or a particular colour or coat. Without genetic testing of the parents, the crosses can end up inheriting genetic defects that occur in both parental breeds. A breed is a group of animals that possesses a set of inherited characteristics that distinguishes it from other animals within the same species. Deliberately crossing two or more breeds is also a manner of establishing new breeds, but it is only a breed when offspring will reliably demonstrate that particular set of characteristics and qualities.

Breed Popularity

Breed popularity varies widely over time and in different parts of the world and different segments of the population. Counting by American Kennel Club (AKC) registration (not by licensing registration or by United Kennel Club (UKC) registration, which could present different statistics), the Labrador Retriever has been the United States's most commonly registered breed of dog since 1991. However, even within parts of the United States, popularity varies; for example, in 2005 the most-registered breed in New York City was the Poodle while the Yorkshire Terrier was the second-most-registered breed in

Houston. However, animal shelters in many parts of the United States report that the most-commonly available dog for adoption is the American Pit Bull Terrier or pit bull-type mixes, making up as much as 20 per cent of dogs available for adoption, none of which would be registered with the AKC. Two decades ago, in 1983, the AKC's top two registered breeds were the American Cocker Spaniel and the Poodle.

In the United Kingdom, The Kennel Club reports that the most-registered breed from at least 1999 to 2005 was the Labrador Retriever. It rounds out the top three for 1999 to 2005 with the German Shepherd Dog, also popular in the States, and the English Cocker Spaniel, which is no longer even in the top ten in the States. In the United Kingdom a national dog adoption and rescue service indicates that the most common breed appearing in shelters is the Greyhound followed by the Staffordshire Bull Terrier.

Physical Characteristics

Modern dog breeds show more variation in size, appearance, and behaviour than any other domestic animal. Within the range of extremes, dogs generally share attributes with their wild ancestors, the wolves. Dogs are predators and scavengers, possessing sharp teeth and strong jaws for attacking, holding, and tearing their food. Although selective breeding has changed the appearance of many breeds, all dogs retain basic traits from their distant ancestors. Like many other predatory mammals, the dog has powerful muscles, fused wristbones, a cardiovascular system that supports both sprinting and endurance, and teeth for catching and tearing. Compared to the bone structure of the human foot, technically, dogs walk on their toes.

Differences from Other Canids

Compared to equally sized wolves, dogs tend to have 20 per cent smaller skulls and 10 per cent smaller brains, as well as proportionately smaller teeth than other canid species. Dogs require less calories to function than wolves. Their diet of human refuse in antiquity made the large brains and jaw muscles needed for hunting unnecessary. It is thought by certain experts that the dog's limp ears are a result of atrophy of the jaw muscles. Dogs tend to be poorer than wolves and coyotes at observational learning, being more responsive to instrumental conditioning.

Dogs have thicker skins than similairly sized wolves. Inuits tend to favour dog skin over wolf skin due to the formers resistance to tear. Unlike wolves, but like coyotes, domestic dogs have sweat glands on their pawpads. The paws of a dog are half the size of those of a wolf, and their tails tend to curl upwards, another trait not found in wolves. Feral dogs show little of the complex social structure or dominance heirarchy present in wolf packs.

For dogs, other members of their kind are of no help in locating food items, and are more like competitors. Dogs differ from wolves and most other large canid species by the fact that they do not regurgitate food for their young, nor the young of other dogs in the same territory.

Many dogs, such as this American Water Spaniel, have had their natural hunting instincts suppressed or altered to suit human needs. Like most mammals, dogs are dichromats and have colour vision equivalent to red-green colour blindness in humans. Different breeds of dogs have different eye shapes and dimensions, and they also have different retina configurations. Dogs with long noses have a *"visual streak"* which runs across the width of the retina and gives them a very wide field of excellent vision, while those with short noses have an *"area centralis"*—a central patch with up to three times the density of nerve endings as the *visual streak*—giving them detailed sight much more like a human's.

Some breeds, particularly the sighthounds, have a field of vision up to 270° (compared to 180° for humans), although broad-headed breeds with short noses have a much narrower field of vision, as low as 180°. Dogs detect sounds as low as the 16 to 20 Hz frequency range (compared to 20 to 70 Hz for humans) and above 45 kHz (compared to 13 to 20 kHz for humans), and in addition have a degree of ear mobility that helps them to rapidly pinpoint the exact location of a sound.

Eighteen or more muscles can tilt, rotate and raise or lower a dog's ear. Additionally, a dog can identify a sound's location much faster than a human can, as well as hear sounds up to four times the distance that humans are able to. Those with more natural ear shapes, like those of wild canids like the fox, generally hear better than those with the floppier ears of many domesticated species.

Dogs have nearly 220 million smell-sensitive cells over an area about the size of a pocket handkerchief (compared to 5 million over an area the size of a postage stamp for humans). Some breeds have been selectively bred for excellence in detecting scents, even compared to their canine brethren. What information a dog actually detects when he is scenting is not perfectly understood; although once a matter of debate, it now seems to be well established that dogs can distinguish two different types of scents when trailing, an air scent from some person or thing that has recently passed by, as well as a ground scent that remains detectable for a much longer period. The characteristics and behaviour of these two types of scent trail would seem, after some thought, to be quite different, the air scent being intermittent but perhaps less obscured by competing scents, whereas the ground scent would be relatively permanent with respect to careful and repetitive search by the dog, but would seem to be much more contaminated with other scents.

In any event, it is established by those who train tracking dogs that it is impossible to teach the dog how to track any better than it does naturally; the object instead is to motivate it properly, and teach it to maintain focus on a single track and ignore any others that might otherwise seem of greater interest to an untrained dog. An intensive search for a scent, for instance searching a ship for contraband, can actually be very fatiguing for a dog, and the dog must be motivated to continue this hard work for a long period of time. Domestic dogs often display the remnants of counter-shading, a common natural camouflage pattern. The general theory of countershading is that an animal that is lit from above will appear lighter on its upper half and darker on its lower half where it will usually be in its own shade. This is a pattern that predators can learn to watch for. A countershaded animal will have dark colouring on its upper surfaces and light colouring below.

This reduces the general visibility of the animal. One reminder of this pattern is that many breeds will have the occasional "blaze", stripe, or "star" of white fur on their chest or undersides. Dogs can generate large amounts of energy for a short period of time. A dog's heart and lungs are oversized relative to its body and its normal everyday needs.

A dog also has relatively more red blood cells than a human. Most of the time the dog will keep the extra red blood cells stored in its spleen. When the animal enters into a situation where its full metabolism is required, such as play, catching game, or fighting other dogs, the extra cells are released into the bloodstream.

The "oversized" heart and lungs will now be running at full capacity, and the animal will have an enhanced ability to engage in aerobic activity. This activity will produce internal heating. Dogs, being covered in fur, are limited in their ability to cool down. After a short time the animal must either cease its athletic activity or risk harming itself from overheating. One can easily observe this pattern of intense activity followed by rest periods in puppies. During the rest phase the spleen collects red blood cells and the animal may pant to cool down.

Dogs are very social animals, but their personality and behaviour vary with breed as well as how they are treated by their owners and others who come in contact with them. It is not uncommon for dogs to attack humans and other animals; however, this is usually because of lack of care or improper upbringing by its owner. Many dogs can be trained to skillfully perform tasks not natural to canines, such as in this dog agility competition. Dogs are valued for their intelligence. This intelligence is expressed differently with different breeds and individuals, however. For example, Border Collies are noted for their ability to learn commands, while other breeds may not be so motivated towards obedience, but instead show their cleverness in devising ways to steal food or escape from a yard. Being highly adaptable animals themselves, dogs have learned to do many jobs as required by humans over the generations.

Dogs are employed in various roles across the globe, proving invaluable assets in areas such as search-and-rescue; law enforcement (including attack dogs, sniffer dogs and tracking dogs); guards for livestock, people or property; herding; Arctic exploration sled-pullers; guiding the blind and acting as a pair of ears for the deaf; assisting with hunting, and a great many other roles which they may be trained to assume. Most dogs rarely have to deal with complex tasks and are unlikely to learn relatively complicated activities (such as opening doors) unaided. Some dogs (such as guide dogs for the visually impaired) are specially trained to recognize and avoid dangerous situations. The meaning of "intelligence" in general, not only in reference to dogs, is hard to define. Some tests measure problem-solving abilities and others test the ability to learn in comparison to others of the same age. Defining it for dogs is just as difficult. It is likely that dogs do not have the ability to premeditate an action to solve a problem.

For example, the ability to learn quickly could be a sign of intelligence. Conversely it could be interpreted as a sign of a desire to please. In contrast, some dogs who do not learn very quickly may have other talents. An example is breeds that are not particularly interested in pleasing their owners, such as Siberian Huskies. Huskies are often fascinated with the myriad of possibilities for escaping from yards, catching small animals, and often figuring out on their own numerous inventive ways of doing both. Assistance dogs are also required to be obedient at all times. This means they must learn a tremendous number of commands, understand how to act in a large variety of situations, and recognize threats to their human companion, some of which they might never before have encountered. Many owners of livestock guardian breeds believe that breeds like the Great Pyrenees or the Kuvasz are not easily trained because their stubborn nature prevents them from seeing the point of such commands as "sit" or "down".

Hounds may also suffer from this type of ranking. These dogs are bred to have more of a "pack" mentality with other dogs and less reliance on a master's direct commands. While they may not have the same kind of intelligence as a Border Collie, they were not bred to learn and obey commands quickly, but to think for themselves while trailing game. Dogs are highly social animals. This can account for their trainability, playfulness, and ability to fit into human households and social situations. These attributes have earned dogs a unique position in the realm of interspecies relationships despite being one of the most effective, voracious, and potentially dangerous predators.

Dogs and humans at times co-operate in some of the most effective hunting in the animal world; in that context, dogs are superpredators. The loyalty and devotion that dogs demonstrate as part of their natural instincts as pack animals closely mimics the human idea of love and friendship, leading many dog owners to view their pets as full-fledged family members. Conversely, dogs seem to view their human companions as members of their pack, and make few, if any, distinctions between their owners and fellow dogs.

Dogs fill a variety of roles in human society and are often trained as working dogs. For dogs that do not have traditional jobs, a wide range of dog sports provide the opportunity to exhibit their natural skills. In many countries, the most common and perhaps most important role of dogs is as companions. Dogs have lived and worked with humans in so many roles that their loyalty has earned them the unique sobriquet "man's best friend". However, some cultures consider dogs to be unclean. In some parts of the world, dogs are raised as livestock to produce dog meat for human consumption. In many places, consumption of dog meat is discouraged by social convention or cultural taboo.

Laughter might not be confined or unique to humans, despite Aristotle's observation that "only the human animal laughs". The differences between chimpanzee and human laughter may be the result of the adaptations that enable human speech. However, some behavioural psychologists argue that self-awareness of one's situation, or the ability to identify with somebody else's predicament, are prerequisites for laughter, so animals are not really laughing in the same way that humans do. The dog-laugh sounds similar to a normal pant. However by analyzing the pant using a spectrograph, this pant varies with bursts of frequencies, resulting in a laugh.

PATENTING OF GENETIC INNOVATIONS IN ANIMAL BREEDING

While herdsmen have practiced the art of animal breeding since the beginning of domestication of livestock, the science of animal breeding is just now entering its second century. From its roots in Mendelian inheritance to the developments in quantitative genetics our field has now grown to include modern molecular sciences. The practical applications of the field have been to genetically improve the animals that produce meat, milk and other products of economic value.

Animal breeding as commerce has always "protected" its endeavors commercially by developing breed societies and using pedigrees to protect the intellectual property developed by the master breeders. The advent of molecular biology, sequencing of genomes and the development of cloning, have brought with them large additional investments. These developments have come with the expectation of increased rewards by investing companies and the reality that technical improvements must be protected in order to obtain sufficient return. The purpose of this paper is to discuss forms of intellectual property, in particular patents and patent law, and to suggest ways it might impact the field of animal breeding and genetics.

Forms of Intellectual Property

Intellectual property (IP) is the property resulting from the creative or "inventive" work of an individual or individuals. The protection of IP is described in a large body of law that includes copyrights, trademarks, trade secrets, and patents. Copyrights protect IP such as articles, books, web pages, computer software and music. Copyright protection applies when the work is in a publishable form. The ã symbol is not required, but recommended and enforcement of rights is made easier by federally registering the mark. In trademark law, the Ô claims common law trademark rights, while the ® indicates federal registration. Trademarks are used primarily to identify the source of the owners' goods, products or services. Examples include "Le Label Rouge", "American Berkshire Gold" and the gene marker *HAL*-1843 ä that is associated with the gene test for porcine stress syndrome. A trade secret represents IP that is not divulged by the owner and confers some competitive advantage. Trade secrets can last as long as the information is kept secret. An example of a trade secret might be specific information to make specialized synthetic lines or calculate breeding values. Trade secret law (in the US called the Uniform Trade Secret Acts) helps protect a technology from theft but the degree of protection is dependent on the type of technology and the means used to keep the information secret.

Patents represent the largest form of IP and their development is rooted in a large body of law both in the US and abroad. Such laws and their interpretation are subject to change, especially when new technology is introduced. The descriptions that follow are meant to be general in nature. Patents were developed to "promote the progress of science and useful arts, by securing for limited times, for authors and inventors, the exclusive rights to their respective writings and discoveries" in return for making details of inventions available to the public. Patenting was developed in order to support the development of science and innovation. A patent, once granted, represents a legal monopoly granted by the respective country's government to an inventor (specific to that country), permitting the patent owner to prohibit anyone else from making, using, or selling this invention for a specific period of time (now generally 20 years). Infringement of a patent is a civil wrong and the owner may sue for economic damages or injunction to stop infringement. Patents are intended to protect but not withhold technical information. Patents are very different from published papers. Inventors, unlike coauthors, must take part in the inventive process, not just participate in the research. Patents are not peer-reviewed but instead are assessed by patent examiners to determine the application's patentability.

A patent application must disclose sufficient detail to enable "one of reasonable skill in the art" to duplicate the invention. The application must be unobvious, *i.e.,* not a simple, obvious extension of existing technology. Novelty is the next criteria and refers to something created that is new. Lastly, the invention must be useful. The scope of patent rights is determined by the claims. Some patents, like gene marker patents, may enjoy very wide coverage across species while others may be limited to single polymorphisms in one breed. Patents may be related to a process, a product produced by a process or dependent on another patent. Patents are open to interpretation by both examiners and lawyers. In the US, patents can be applied for up to one year after publication or public disclosure but this is not allowed in other countries. Patents in and of themselves do not ensure income. They must be promoted and protected and can be viewed as tradable assets and licensed or assigned to third parties. Confusion often exists relative to "international" patents, which actually do not exist. Under the Paris Convention Treaty, a group of countries agreed to work with each other to provide a unified examination process and to make filing in its member countries easier by forming the World Intellectual Property Office (WIPO). In unison they adopted the one-year period in which to file an application in one of the other countries without losing the benefit of the original filing date. Using this procedure avoids the citation against the applicant of any "art" that became known after the original filing date in the country of origin, but before the filing date in another country. Under this Patent Cooperation Treaty (PCT), an applicant is allowed one year to file at the WIPO office and, by designating member countries, holds the legal rights and original filing date in those designated countries without having to file in each country or pay the national expense up front. Although there are fees associated with the PCT application, they are less than one would pay for the foreign national filings at a later date. To obtain a patent in these countries, the application does eventually need to be filed in the national patent offices, fees paid, translations done and the laws and regulations of each country office followed. The applicant may have additional time from the national filing date to request examination, depending on particular country requirements. Delaying examination may provide the time desired to continue development and/or commercialization, however, during this period, fees called "annuities" must generally be paid. Due to the expense involved, applications are generally filed only in countries where protection is really needed.

A Brief History of Patents

The early protection of intellectual property rights can be traced back to Venice in the Middle Ages when master craftsmen in guilds prohibited competition from former apprentices for a period of 20 years (Waltersheid, 1994). Such laws had considerably different economic effects for the master craftsman, the apprentice and the general public. For livestock, early breed societies were developed and monitored pedigrees to protect the IP of the master breeders.

The first modern patent act is often thought of as The US Patent Act of 1790. There was similar legislation in France in 1791. Patents related to living matter are relatively new. One of the earliest patents for living matter was granted to L. Pasteur for a yeast strain but this was done under the belief that it was an inanimate object and not living (Lesser, 2002). The first specialized patent law applied to living organisms was that of the Plant Protection Act of 1930 in the US and provided what is commonly referred to as Plant Breeders Rights (PBR) to propagate new varieties by asexual methods. In 1961, a similar law was passed in France called the UPOV (International Union for the Protection of New Varieties of Plants - Union Internationale pour la Protection des Obtentions Végétales; http://www.upov.int). Protection in the US was expanded in 1970 with the Plant Variety Protection Act to include sexually reproduced plants. The UPOV was revised in Europe in 1991. Under these laws two principles, "breeder's rights" which allows breeders to use protected varieties without permission of the owner and "farmer's privilege" which allows farmers to collect seeds from their crops and use them, were developed (Lesser, 2002). For years many seed companies have attempted to halt this latter practice by asking farmers to sign contracts prohibiting it. Recent technology, like the "terminator" technology (Kaiser, 2000), biologically prohibits the practice. Santaniello *et al.* (2000) provides an excellent review on all aspects of patent protection for plants.

The dawning and blossoming of the molecular age presented real problems for the protection of IP related to living organisms other than plants. In 1975, a French company failed to patent a "dwarf, egg-laying chicken hen produced by a process" involving a sex-linked recessive gene (Bent, 1989). In 1980, the United States Supreme Court in a 5-4 decision (*Diamond v. Chakrabarty*) declared that "anything under the sun that is made by man" is patentable. This case concerned the patenting of genetically engineered bacteria that destroyed oil sludge. In 1987, the US Patent trade office issued a pronouncement of the patentability, in principle, of non-human multicellular organisms that were not naturally occurring (Bent, 1989). This was quickly followed in 1988 by the landmark patent on the so-called "Harvard mouse" which was engineered to be susceptible to cancer. In Europe, similar laws were passed allowing patenting of animals and animal-related inventions. From 1995 to 2001, a total of 45 animal patents were granted in the US (Lesser, 2002). While these changes in patent law had large economic implications, the social and moral ramifications were also enormous.

ANIMAL BREEDING AND GENETICS - IP PROTECTION

The broad area of biotechnology encompasses many of the patent applications in our field and the US Supreme Court has established guidelines that apply to this technology (Nebel *et al.*, 2002). The Court made it clear in *Brenner v. Manson* that patent utility implied usefulness and not just "any invention not positively harmful to society." The Court expressed reservations regarding a monopoly on compounds with unknown functions, and that utility must extend beyond proving that the product is a result of scientific research (Nebel *et al.*, 2002).

Technologies in the field of animal breeding that may require (or qualify for) IP protection include, but are not limited to: statistical methods for genetic improvement, DNA markers for genetic improvement, transgenic and cloned animals and methods to produce them, new methods to measure traits, methods to identify animals, computer software and other written materials.

Patent coverage in not just confined to genetic markers. Lines of pigs or chickens have been patented and can be viewed as a specialized extension of early breed development or trademarking for protection of this IP. Other patents exist for methods involving cellular and animal manipulation and involve processes like stem cell development, transgenic production (*i.e.*, US 6,271,436) and cloning (*i.e.*, US 6,215,041 or US 6,258,998). Several advances related to mechanical or electronic devices have been made and include new A.I. or embryo transfer tools, advanced ultra-sound equipment, formulas and methods to measure backfat and other traits in livestock. The increasing need for traceability of animals and animal products has spawned a number of inventions including electronic ID tags and retinal scanning methods and devices.

Considerable discussion has ensued recently from a patent entitled "Method of Bovine Herd Management" granted in the U.S. in 1994 and later in Canada (Schaeffer, 2002). The invention is for the "test-day model" and includes the gathering, mathematical treatment, and the use of the modified data by dairy producers. The novelty and unobviousness of the patent has been seriously questioned. It was pointed out that the practices of gathering, manipulating and using data by dairy producers have existed for nearly 100 years. The patent therefore claims rights to a practice that has been public knowledge for a long time. The novel idea within the patent was the specific mathematical model and procedures that Everett and co-workers developed for the analysis of test day yields.

Everett was also not the first researcher to apply a model to test day records and as has been demonstrated, the model as described in the patent, is not necessarily the best model that could be applied (Schaeffer, 2002). This patent generated considerable discussion. It has been argued "what would the field of animal breeding be if the selection index or Henderson's BLUP had been patented?" Yet while quantitative geneticists see the thought of such protection as sacrilege, molecular scientists accept (but may not like) that in a similar way the foundation patent for PCR exists and royalties must be paid for its use.

Other Legal Considerations

In the US and some other countries, there is no clear exception to infringement for research use of patented inventions, even in non-profit or educational institutions. Researchers are required to obtain a license or permission from the patent owner to use a protected invention for research purposes. In other countries, there is either an exemption for educational or non-commercial research or research on improving the invention. Research using the invention, however, is not allowed. Researchers should also be concerned with possible claims for "contributory infringement," *i.e.*, assisting someone else to infringe on a patent. For example, due to a claim of possible contributory infringement, the US Genome Coordinators were asked to halt the practice of supplying primers for microsatellite analysis to other researchers. To resolve the issue the Coordinators began buying the primers from a company with a research license.

Ethical, Social and Economic Issues

Many ethical and social issues have been raised related to patenting of animals and genes (Bent, 1989; Brody, 1989; Dresser, 1988; Evans, 2002). These include: 1) patenting of animals or genes will be destructive to nature and allows man to play "God"; 2) patenting will devalue animal life and hence human life; 3) patenting will increase animal suffering; 4) patenting will lead to a decline in genetic diversity of animals and threaten species; 5) patenting speeds the trend towards commercialization of academic research and 6) patenting will undermine conventional farming and lead to increased industrial farming systems. Early humans domesticated animals and master breeders and geneticists have transformed them into productive species. Was this playing "God" or interfering with nature? The use of transgenics for making specialized animal lines for disease research is certainly adapting nature but does it devalue life and is it unethical or immoral? These are value judgements that most in society have decided are acceptable (Brody, 1989). Certainly some lines have been (and need to be) drawn to delineate what is acceptable and unacceptable. For example, most people and governments have concluded that while cloning of animals is acceptable (at least for research purposes), cloning of humans, for whatever reason, is immoral, unethical and to be avoided.

Animal welfare and animal rights issues continue to be at the forefront of livestock production and biomedical research (Evans, 2002). Individuals who believe that animals have "rights" will likely be opposed to patenting any invention derived from animal research. The most commonly cited examples are those relating to transgenic animals (*i.e.*, early transgenic pigs) in which some animals had health problems. Production and patenting of specialized lines of rodents for biomedical research that have a tendency to develop specific diseases is also considered unethical by animal rights activists (Brody, 1989). If, however, individuals believe that animal rights are subordinate to those of humans, but that they deserve proper care and welfare then the issue of patenting is much less of a concern.

It has been suggested that through the use of gene markers and highly selective breeding, or through the use of transgenics and cloning, that genetic diversity will be greatly reduced. Certainly these methods have the potential to remove within-line variation but they are likely to increase between-line variability.

However, it may be argued that the patent system in fact encourages diversity as it promotes and helps establish, via patent-related deposits of biological materials, a broader genetic diversity (Bent, 1989). Related to this issue is the issue of patenting of products from animals or plants from developing countries.

Should companies be allowed to sample and alter wild stocks and to subsequently earn great sums of money by then selling them to both developed countries and back to the countries from where they were obtained? Some individuals say this so-called biopiracy is encouraged by patenting and some countries are moving to address these concerns (Evans, 2002). An extension of this is the issue of species integrity. The development of transgenic pigs for xenotransplantation and other species with human genes for human protein production threatens the genetic lines that determine species. While consideration of xenotransplantation has slowed due to fears of retroviruses and diseases like mad cow disease and AIDS, as Evans (2002) points out this dispute over species integrity "is not empirical" but it is tied up in the individual and societal views on nature and "the notion of awe and wonder."

It is likely, no matter how repugnant this process is to some, there will be development of transgenic lines of animals for biomedical research (not food production) and applications that do encompass genes from other species.

In a perfect world, public support of research would be 100 per cent of required funding and all IP would be publicly available in the country of origin. Many US public institutions are now funded at about 40 per cent of total budgets and pressures to obtain outside funding are growing. Companies supplying funding expect to "own", through licensing agreements or otherwise, the IP that results. Protection of IP can provide research partners a greater basis for trust and exchange of ideas and help insure that public institutions focus research on areas of high relevance and payoff. Commercial relationships also aid in technology transfer, employment opportunities for students and may allow for research opportunities not available in the public sector. Fretz and MacKenzie (2002) have suggested that: 1) such activities serve the public good; 2) short term, low quality research should not be favoured in order to obtain funding; 3) managing of IP must be a balance of serving the institution, funding agency and public; 4) independence of the researcher and institution must be maintained; 5) the mission of the public institution is not altered and 6) conflicts of interest should be avoided. Institutions must be proactive in balancing funding, IP protection and commercialization.

Does patenting increase the likelihood of "industrial farming?" Economic and governmental issues certainly play a role in the size of the average farm and the level of commercialization of farming. It can be argued that if large companies have exclusive licensing arrangements for genetic tools then small breeders will be disadvantaged. Market pressures related to size of operation and efficiency of production are much larger influences on the industrialization of farming and livestock production than patenting.

Other issues regarding patenting that have an economic basis can be discussed. Langinier and Moschini (2002) provide an excellent review of the economics of patenting and the benefits and costs that are derived. While inventors would prefer broader claims, limited claims encourage competition and further innovation. Complex products often require building on other patents. Unfortunately, development does become blocked by other patents (especially at the leading edge) and this has negative effects on both the developer as well as the public. Licensing of patents exclusively to one company may benefit that company and segments of the public but might also limit further innovation and not be in all the public's best interest. Langinier and Moschini (2002) concluded "that continued efforts are required to improve the workings of the patent system" to improve the economic performance of the system.

SELECTIVE BREEDING

Selective breeding in domesticated animals is the process of a breeder developing a cultivated breed over time, and selecting qualities within individuals of the breed that will be best to pass on to the next generation. Breeding techniques such as inbreeding, linebreeding and outcrossing are utilized by breeders in the maintenance and improvement of their chosen breeds.

Breeding Stock

"Breeding stock" is a term used to describe a group of animals used for purpose of planned breeding. When an individual is looking to breed animals, he or she is looking for certain valuable traits in purebred stock for a certain purpose, or may intend to use some type of crossbreeding to produce a new type of stock with different, and presumably superior abilities in a given area of endeavor.

To breed chickens, a typical breeder intends to receive eggs, meat, and new, young birds for further reproduction. Thus the breeder has to study different breeds and types of chickens and analyse what can be expected from a certain set of characteristics before he or she starts breeding them. Accordingly, when purchasing initial breeding stock, the breeder seeks a group of birds that will most closely fit the purpose intended.

Purebred Breeding

The idea of *breed purity* may strike an unpleasant chord with some individuals because it is reminiscent of nineteenth-century eugenics notions of a "superior strain" in humans, supposedly exemplified by members of an aristocratic social class or of certain races. The application of theories of eugenics had far-reaching consequences for human beings, particularly in cases such as The Holocaust. However, the idea of a superior strain in animals, such as that incorporated in the breeding of Thoroughbred race horses, was less controversial. By "breeding the best to the best," employing a certain degree of careful inbreeding, considerable culling, and selection for "superior" qualities, one could develop a bloodline or "breed" superior in certain respects to the original base stock or landrace which had been produced primarily by natural selection.

This process forms the basis for a "true-breeding" line of animals, which today is generally called a "purebred" line. Once created, a purified line is "closed" and preserved from dilution and possible "debasement" by unrelated stock. However, this theory is not without some negative consequences. When the ideal of purified lineage or purely aesthetic type is seen as an end in itself, without consideration of usefulness or vigour, a breed can suffer over time. Also, if a breed registry is closed with too small a gene pool, over time, degradation due to inbreeding depression may occur. For many purebred breeders, the need for a periodic outcross is recognized and incorporated into an overall plan. The observable phenomenon of hybrid vigour stands in contrast to the notion of breed purity. However, on the other hand, indiscriminate breeding of crossbred or hybrid animals may also result in degradation of quality, particularly when animals are mass-produced without concern for overall health or longevity.

Backyard Breeding

The term backyard breeder is a general term, sometimes considered derisive, used to describe people who are "home breeders", and breed only a few animals for personal use or occasional sale. However, while some backyard breeders pay strict attention to quality, the term usually describes those who allow animals, particularly dogs or horses, to procreate regardless of physical, genetic, and/or emotional health. It is also used to describe unreputable commercial breeders. In the process of careless breeding, many backyard breeders produce genetically weak animals that can be predisposed to genetic disorders or debilitating physical deformities, particularly those that result from inbreeding.

In dog breeding, these individuals focus on only a limited aspect of the dog, while ignoring the original function and temperament of the breed.

When such breeding is carried out on a large scale, the venue is called a *puppy mill* or *puppy farm*. Backyard breeding is popularly blamed for the proliferation of aggressive dogs for the sports of dog baiting and dog fighting.

Dog fanciers generally believe that such ill-bred dogs are the reason for the bad public reputation of certain dog breeds, and the increasingly frequent enactment of breed-specific legislation. Because of the time and expense of feeding and caring for horses, which produce one foal per year, there are fewer horse breeders who produce animals *en masse,* though some individuals do engage in animal hoarding and breed far more animals than they can support.

In the horse world, over breeding of grade animals that cannot be sold except for very low prices raises concerns that such animals will be slaughtered for horsemeat. Some, though not all, colour breeders make errors similar to puppy mill breeders by focusing on only one characteristic to the detriment of health, soundness, and quality conformation. In addition, industrial breeders, such as those who produce *pregnant mare urine* for the manufacture of premarin, allow many mares to be bred indiscriminately, simply so that they are able to produce the urine laced with reproductive hormones that forms the base for the drug. Their foals are generally considered an unwanted byproduct, particularly males (colts), and sold off, frequently to the slaughter market.

IN-BREEDING AND OTHER BREEDING METHODS

One of the most hotly talked about topics with regard to pure-bred dogs is the use of in-breeding. This is a term that is often misused and is extremely misunderstood. Part of the misunderstandings comes from differences in the way the terms are used within the scientific/medical field, and how it is commonly used by breeders. These are the most commonly accepted definitions used by serious dog breeders and will be the definitions used within this article.

- *In-breeding:* This is the breeding of closely related animals. Brother-Sister, Parent-Offspring.
- *Line-breeding:* This is the breeding of animals that share common ancestors but are not closely related. The dogs may share a common great-grandparent.
- *Out-cross:* This is generally considered the breeding of animals with no common ancestors within the first 4 or 5 generations.

Common Misconceptions

- *In-breeding causes genetic diseases:* Breeding closely related animals increases the possibility that any bad genes in a line will show up. It does not 'cause' genetic disease.
- *Out-crossed dogs are healthier:* This is only partly true. There is a known phenomenon called Hybrid Vigour. Two animals of unrelated strains breed and the offspring is often bigger and grows faster than it's purebred cousins. This method is often used by farmers in order to get their animals to market sooner. But one of the biggest misconceptions of hybrid vigour is that it applies to all animals of mixed heritage. Hybrid Vigour only applies to the animals that are the direct offspring of the crossing of the unrelated strains. In other words if you continue to breed animals of different strains there generally will not be any additional increase in hybrid vigour. If the unrelated strains share common genes for genetic disorders, hybrid vigour will not over ride the risk of the disorder showing up. Out-crossing can also cause problems if widely divergent physical types are mixed due to differences in growth rates and bone and muscle sizes.

BENEFITS OF EACH TYPE OF BREEDING

By definition, purebred dogs have a smaller gene pool to draw on than mixed breed dogs. That smaller gene pool gives the breed its individual characteristics, such as physical appearance and temperament. It is what makes a poodle a Poodle and a Golden Retriever a Golden Retriever. But there is considerable controversy with regard to whether the gene pools of the modern pure-bred have become too small.

Inbreeding

In-breeding is more likely to help "set" or "fix" a particular trait within a breed or a line by narrowing the gene pool to favour those traits. So if a breeder is looking to set a particular desirable feature of their line then in-breeding and choosing the offspring most strongly possessing that trait can be beneficial. In the first pedigree Sydney is the maternal son of Annie, thus doubling on Annie's genes. Or another way to look at it is Annie is both Sydney's mother and Grand-mother. In the second example we have a full brother/sister in-breeding so we are doubling up on the genes from both Kiley and Annie.

In-breeding can also help identify those bad genes that exist within a line. Dogs possessing the bad genes can be eliminated from a breeding programme and carriers also identified. Intermittent in-breeding within a line or breed is not damaging to the long term health of the animals. However, in-breeding over successive generations can lead to reduced fitness and fertility problems among the offspring, resulting in phenomena known as In-breeding Depression.

It can take many generations to show up depending on the traits involved. To use this method responsibly a breeder would not want to in-breed on animals with known genetic disorders, temperaments not in keeping with its given breed, or known serious structural faults, or to in-breed frequently even on healthy-superior specimens.

Line-Breeding

Line-breeding is another way to help "set" or "fix" desirable traits. With line-breeding we breed animals that are related, but we are also routinely introducing genes from other lines into the genetic mix. It takes longer to fix the desirable traits this way, but doing so lowers the risk of those problems associated with repeated in-breeding.

Loose line-breeding over successive generations will result in more variations of physical appearance than would in-breeding or tight line-breeding, but will keep the physical look and structure within the same general size and shape, it also carries fewer long term risks.

According to geneticists: Line-breeding can be carried on for many many generations without deleterious effects on the line or breed as long as the individuals involved have few hidden genetic disorders.

Out-Crossing

Out-crossing in terms of pure-bred dogs is the breeding of unrelated dogs. On a pedigree no names will be repeated within the first 5 generations. This type of breeding has both advantages and disadvantages. Which as it turns out are flip sides of the same argument. The greatest genetic diversity are maintained by out -ccrossing, but this also leads to the least consistency in terms of physical appearance and other traits.

Out-crossing does not guarantee that the animals won't develop genetic disorders, but it does tend to reduce the numbers of affected offspring. Best chance of getting an animal that is less prone to developing a genetic disorder comes more from finding a conscientious breeder that screens their animals for hereditary disorders and breeds for the betterment of the breed. All three methods of breeding have their place in a long term, well thought out breeding programme. Talk to the breeder, ask questions as to what their goal is in doing a particular breeding. Ask about the risks and what problems are known to that line. And all lines have some because the perfect dog and the perfect lines are still goals of the future for all breeders.

SELECTION AND EXPECTED RESPONSE

Traits which are heritable presumably can be changed by selection, but the degree of heritability is only one of three factors which determine the progress which can be made. The other two are the culling level (intensity of selection) and the variability of the trait. These are logical since the more severely we can cull, the more progress we can make. Also, the more variable a trait is, the higher the population average will be for the selected individuals for any given culling percentage.

It turns out that the amount of progress which can be made in one generation is the product of intensity of selection, heritability for that trait, and the observed variability. Note that if any of the three components is zero, selection progress will be zero also. Good estimates are available for how much progress can be made from selection for most of the economically important characteristics of dairy cattle. Of course, there are several traits which are important, so there is a need to look further than anticipated results from selection for a single trait.

Correlated Response

Knowing what the genetic relationships are among the important traits, plus a few other things, permits us to predict what will happen to a trait such as fat percentage if we ignore it completely and select for milk yield, or vice versa. Though these results pertain to Holsteins, they are similar to results for other breeds. The changes expected in the trait being selected for; values off the diagonal are changes expected to happen in other traits. Heritabilities and variabilities used in these calculations agreed with those generally accepted by dairy cattle geneticists.

Intensity of selection in the estimations was fairly conservative. If more severe culling were practiced, greater progress would be made. If milk yield alone is selected for at a reasonable level of intensity, an increase is expected in genetic merit in the next generation of 607 pounds. Since milk yield, fat yield and protein yield are genetically fairly closely related (that is, their genetic correlations are high and positive), an increase is expected in fat and protein yields, even though these were ignored in the selection programme.

Because of the negative genetic correlations between milk yield and fat percentage, and milk yield and protein percentage, these two percentages are expected to go down. Selecting for milk yield and ignoring the other four traits would be expected to result in an increase per generation of 607 pounds of milk, 23.2 pounds of fat and 13.7 pounds of protein, but decreases of 0.036 per cent fat and 0.018 per cent protein. If one attempted to increase fat percentage genetically in his herd by selecting for it and ignoring other traits, he could increase fat percentage. A reasonable increase per generation to be expected is 0.190 per cent. Along with this, however, there would be an actual decrease in milk yield of 287 pounds, although fat yield would increase 24.1 pounds.

Present pricing systems for milk in the United States are based mostly on its weight and fat percentage, although additional factors such as protein percentage may be included also. So the dairyman needs to decide what combination of milk yield and fat percentage and other changes is most profitable for him. He also needs to produce milk that passes minimum legal standards; this is an important factor in herds which already are near the minimum. Beyond this, there are a number of other important traits which must be considered. Dairymen, of course, mentally put all of these factors together in their selection programmes and, depending on their present herd status and present milk pricing, give each trait a certain amount of emphasis. In other words, they weight each characteristic they think is important. Geneticists do the same thing and attempt to establish numerical values for the weights; a numerical index, called a selection index, then can be formulated for each economic situation.

Milk Yield and Composition

Just about every index formulated for the United States gives yield more emphasis than any other trait, with enough emphasis placed on fat percentage to meet legal standards. Protein has become more important in pricing systems in recent years.

Reproductive Performance

Efficient reproduction is very important economically because a cow must first reproduce to produce. However, most measures of heritability are zero or close to zero, suggesting that little or no improvement can be made by selection. There already is considerable selection pressure being exerted on improving reproductive performance without any conscious effort being made by the breeder.

This occurs because sterile males and females leave no offspring, and those with low efficiency leave a lot fewer than those of higher reproductive performance. So within a given herd culling for reproductive failure would have little effect on the genetic merit for reproduction of the herd. Now that large numbers of records are being obtained on reproduction of daughters of AI sires, however, it may turn out that this information can be taken advantage of in selecting which proven AI sires to use in the herd. Even though heritability of most single measures of reproduction are close to zero, estimates of average genetic merit of a large group of daughters for reproductive efficiency are much more reliable.

Desirable Body Type

Desirable type has appealed to dairymen for centuries, and major changes in appearance in dairy cattle have occurred over the years, accompanied by increases in milk yield. This fact does not itself mean that the two traits have (or ever have had) an appreciable, positive genetic correlation. Regardless of what has gone before, the breeder must concern himself with the present facts of life in his selection programme. The genetic relationship between type and production seems close to zero and may even be negative. There have been a number of changes in the breed association type classification programmes and policies in recent years, and evaluation of all such changes cannot yet be made. A number of scoring systems also have been developed by others such as universities and AI studs. Yet the final value of desirable type must be in its relationship with production and longevity of the cow.

Except for the breeder who derives an important fraction of his income from sale of breeding stock, dairymen should realise (according to research) that only 2 to 6 per cent of their income is related to variability in body confirmation. USDA researchers estimated several years ago that giving equal emphasis to type and production would reduce the possible increase in milk yield by 29 per cent. Though perhaps the final judgement on emphasis to be placed on desirable type cannot yet be made, it appears that if major attempts are made to improve type, they will be made at the sacrifice of possible genetic increases in milk yield.

Longevity

Longevity may be defined as the ability to avoid death or culling, the former being mostly for disease or accident and the latter for reproductive failure, low milk yield, mastitis, etc. Few cows die simply from old age. Of the various possible reasons for loss, some have a genetic basis but few are highly heritable. In any event, high producing cows remain in the herd longer than low producers. Natural selection pressures also work in favour of the ability to resist death or culling. Selection for high milk yield, plus natural selection, seem to be exerting sufficient selection pressure for increased longevity such that the breeder need not give it major emphasis in his programme.

Body Size and Weight

Within each dairy breed, large cows give more milk, on the average, than small cows, but they also require more feed. Research suggests that the additional costs involved with larger cows about equal the additional income derived from them. In the case of two cows of equal production but different size, one actually would expect the smaller cow to be more profitable. Most measures of size (weight, height, etc.) show appreciable heritability and variability, so larger cows could doubtless be developed if desired.

However, most research now suggests that milk yield in dairy cows can be increased considerably without there being any increase in size. Selection for increased body size could be successful, but the correlated response in milk yield though probably positive would be close to zero. Though North Carolina researchers indicated that it might still be wise to check cows considerably above or below the breed average in body weight for productivity and longevity, they concluded that "undue emphasis on larger cows in the short run could lead to the development of less profitable animals."

Other Traits

Many others are economically important and are heritable to some degree. Many qualitative (simply inherited) traits, ranging from lethals on one extreme to aesthetic traits on the other, can be important under some circumstances. Artificial insemination, particularly frozen semen, has caused concern because of possible widespread dissemination of undesirable genes. The genetics of the situation have been studied, however, and with present knowledge potential dangers apparently can be avoided. Traits other than those mentioned also are important. Mastitis resistance, is of extreme economic value and has measurable heritability. A certain amount of direct pressure for resistance will be exerted from culling of mastitic animals, although clinical cases may remain in a dairy herd for years under proper treatment.

Present knowledge of genetics of dairy cattle suggests that major selection emphasis should be placed on milk yield, with enough selection pressure on fat percentage to maintain composition of milk at legal standards. Other components of milk may become important if pricing systems change to emphasize them. Care should be taken to include as few traits as practical in a selection programme, since progress in milk yield may be sacrificed to a degree such that total profit per cow may decrease. This is especially true with traits of low heritability and only slight genetic relationship to milk yield. Reproductive efficiency and resistance to mastitis may appear as economically important traits in future sire proving systems.

The Animal Model

The Animal Model does many things notably different from previous evaluation systems. These include simultaneous evaluation of all sires and cows. This means all information on all relatives is used to increase the accuracy of any one animal's evaluation. This also means that cow families are directly included in the evaluation of all sires and cows, effectively eliminating the basis for previous objections by some breeders. All female ancestors and descendants are used in the evaluation of both sires and cows, weighted by how closely they are related. Since all known relatives of both registered and grade cows are used, computing the evaluations for the Holstein breed requires simultaneous solution of about 50 million equations. Other changes include correction for genetic merit of mates, use of a new genetic base, new ways of using records on cows missing first records and cows changing herds, and definition of different management groups for registered and grade cows.

There is some new terminology with the Animal Model emphasizing that new methods are truly different. Predicted Difference (PD) changes to Predicted Transmitting Ability (PTA), or one-half the breeding value. This is the prediction of what the sire transmits to his offspring. Cow Index changes to Predicted Transmitting Ability (PTA).

Repeatability (RPT) changes to reliability. Repeatability always has been a misnomer as used by the USDA in this context anyway. Since reliability includes contributions from additional relatives, reliability figures should be higher than the old repeatabilities. The biggest difference is for sires of cows with few daughters compared to those with many. A sire with no daughters would have a reliability of 37 per cent. per centile is sire's ranking based on the active AI bull population; for cows it will be based on cows with recent lactations. The rankings are based on an index including milk, fat, and protein evaluations. Previously they were based on an index including only milk and fat. As good as the new Animal Model system is, it cannot correct for preferential treatment of daughters of certain sires. No genetic evaluation system can. Random AI sampling of young sires and wide daughter distribution will continue to be important in establishing accurate first evaluations. This also is true for sires with relatively few daughters.

Economic Evaluation of Sires and Their Semen

Predicted transmitting ability for dollar value (PTA PRO$) is based on milk, fat, and protein evaluations. Current national prices are used. For the July 1989 run these were $11.20 per hundredweight of milk containing 3.5 per cent milk fat and 3.2 per cent protein. Hence the equation for estimating PTA PRO$ is PTA PRO$ = $.0242 (PTA milk) + $ 1.54 (PTA fat) + $ 1.09 (PTA protein). Note that the dollar value is computed from pounds fat and protein produced, not percentages. per centile rankings are based on this index.

In Florida, the milk price is higher than almost everywhere else in the United States, and the dollar values of superior sires should be proportionally higher. The ranking of bulls by evaluations would not change much, but the net value over their semen price would. Taking advantage of considerable previous research, Marsha Wilcox and coworkers developed techniques to aid dairymen in the selection of semen to purchase for herd use. The concept is that of net present value (discounted net income) of semen.

Although the method is computationally elaborate, availability of microcomputers to dairymen will permit each to develop a system of evaluating semen based on a set of selection criteria tailored to individual needs. Information necessary for semen selection include sire proof for milk, fat, protein, type (PDT), interest rate, and such management parameters for the individual herd as conception rate, calving interval and female mortality rate. Outcome of this linear net merit index for semen is Present Value Dollars (PV$).

These researchers then evaluated a number of possible economic and management situations to determine the effects on sire selection; considerable shifting of rank occurred with changes in criteria. For one sample selection programme (all selection pressure on milk, none on desirable type), rank correlations of PV$ and PDM, and PV$ and PD$ + semen price, were.56 and.52. Such procedures should prove useful and profitable to dairymen who are able to characterize their own management parameters and to define their selection goals. Thus, the concepts of proving sires are not difficult. Efforts to obtain better estimates, thus increasing rates of genetic improvement, continue to make the exact calculations and formulas very complex.

CATTLE BREEDING

Reproduction is an important consideration in the economics of cattle production. In the absence of regular breeding and calving at the appropriate time cattle rearing will not be profitable. A healthy calf each year is the usual goal. This is possible only by increasing the reproductive efficiency of the animals. Successful reproduction encompasses the ability to mate, the capacity to conceive and to nourish the embryo and deliver the viable young ones at the end of a normal gestation period. In fact, interruption in this chain of events leads to failure of the cow either to conceive or the embryo to die or to have a premature delivery of the foetus.

The reproductive efficiency is a complex phenomenon controlled by both genetic and non-genetic factors, the non- genetic factors being climate, nutrition, and level of management. The reproductive efficiency varies not only between species and breeds but also among the animals within the Same breed. Even the best feeding and management can not coax performance beyond the genetic limit of an inferior animal. Improving the genetic merits of livestock populations is important at all levels of management. A sound breeding programme is a necessary part of the total animal production system. It is absolutely imperative to improve the productive capacity and physical appearance of the animal population.

Factors Affecting Breeding Efficiency

The factors which influence the breeding efficiency of cattle are as follows:

Number of Ova

The first limitation on the breeding efficiency of fertility of an animal is the number of functional ova released during each cycle of ovulation. Ovulation is the process of shedding of ovum from the Graffian follicle. In the case of cow, usually a single ovum is capable of undergoing fertilization only for a period of 5-10 hours. Therefore, the time of mating in relation to ovulation is important for effective fertilization.

Percentage of Fertilization

The second limitation is fertilization of ova. Failure to be fertilized may result from several causes. The spermatozoa may be few or low in vitality. The service may be either too early or too late. so that the sperms and eggs do not meet at the right moment, to result in fertilization.

Embryonic Death

From the time of fertilization till birth, embryonic mortality may occur due to a variety of reasons. Hormone deficiency or imbalance may cause failure of implantation of fertilized ova which die subsequently. Death may occur as a result of lethal genes for which the embryos are homozygous. Other causes may be accidents in development, over-crowding in the uterus, insufficient nutrition or infections in tile uterus.

Age of First Pregnancy

Breeding efficiency may be lowered seriously by increasing the age of first breeding. Females bred at a lower age are likely to appear stunted during the first lactation, but their mature size is affected little by their having been bred early.

Frequency of Pregnancy

The breeding efficiency can be greatly enhanced by lowering the interval between successive pregnancies.

The wise general policy is to breed for the first time at an early age and to rebreed at almost the earliest opportunity after each pregnancy. In this way the lifetime efficiency is increased. Cows can be rebred in 9-12 weeks after parturition.

Longevity

The length of life of the parent is an important part of breeding efficiency, because the return over feed cost is greater in increased length of life.

Also, it affects the possibility of improving the breed. The longer the life of the parents, the smaller the percentage of cows needed for replacement every year.

Management Practices to Improve Breeding Efficiency

Some of the management suggestions which will tend to improve breeding efficiency of cattle are listed below.

1. Keep accurate breeding records of dates of heat, service and parturition. Use records in predicting the dates of heat and observe the females carefully for heat.
2. Breed cows near the end of heat period.
3. Have females with abnormal discharges examined and treated by veterinarian.
4. Call a veterinarian to examine females not settled after three services.
5. Get the females checked for pregnancy at the proper time after breeding.
6. Buy replacements only from healthy herds and test them before putting them in your herd.
7. Have the females give birth in isolation, preferably in a parturition room and clean up and sterilize the area once parturition is over.

8. Follow a programme of disease prevention, test and vaccination for diseases affecting reproduction and vaccinate the animals against such diseases.
9. Practice a general sanitation programme.
10. Supply adequate nutrition.
11. Employ the correct technique.
12. Provide suitable shelter management.
13. Detect silent or mild heat, by using a teaser bull.

Selection and Culling

Selection and culling are the two sides of the same coin. Selection is the process in which certain individuals in a population are included for becoming the parents of the next generation. Automatically some are excluded for the purpose which are culled. Natural selection has been going on since ages where animals which were stronger, which had better survivability and which were in more unison with the environment around them, found a better chance to reproduce. Thus certain genes for certain characters got more chance to be selected to form individuals in the subsequent generations. Since domestication of cattle, man has been looking for superior phenotypes in traits useful to him and selecting such animals to form the parental generation. This is man made artificial selection. Now man has progressed one step further in making estimates of genotypes from the study of phenotypes and making use of that information (in artificial) selection.

Selection Methods

There is only one way to select and that is to "keep the best and cull the poorest. The various selection methods are techniques for identifying or estimating the genetic values of individual candidates for selection. The procedure discussed here apply to selection for quantitative trails.

Performance Testing

Performance test is a measure of the phenotypic value of the individual candidates for selection. Since the phenotypic value is determined by both genetic and environmental influences, the performance test is an estimate, not a measure of the genetic value. The occurrence of this estimate depends upon the heritability of the trait *i.e.* on the degree to which the genetic value is modified by the environmental influences.

Advantages:

a. Among simple procedures, the performance test is the most accurate.
b. Environmental influences can be minimised by testing candidates for selection in the same pen or in similar environmental conditions.

c. The measure is direct, not on a relative basis.
d. All candidates for selection can be tested in contrast to progeny testing where only a parent can be tested.
e. Generation intervals are usually short.
f. Testing can usually be done on the farm under normal management conditions.

Disadvantages:

a. Accuracy become low when heretability is low.
b. Phenotypes are not available for one sex or in sex limited traits such as milk yield.
c. Traits which are not expressed until maturity may become expensive or difficult to manage by performance tests since most selection decisions must be made before maturity.

Performance tests should be the backbone of most selection programmes. Although much publicity has been given to other selection methods, it remains a fact that most of the progress in livestock improvement to date has been due to selection on the individual's Own phenotype *i.e.* performance test.

Pedegree Selection

A pedegree is a record of an individual's ancestors including its parents. This information is valuable because each individual possesses a sample half of the genes from each parent. If we can precisely know an individual's phenotype, little is gained by considering pedegree in selection. Pedegree considerations are useful when we do not have sufficient accurate records of production of the individual.

Also, it is useful in the early selection when the traits in question might not have expressed themselves. It is also useful for selection of males when the traits selected for are expressed only by the female (sex-limited) such as milk production in dairy cattle.

Advantages:

a. It provides information when performance tests are not available for the candidates.
b. It provides information to supplement performance test information.
c. It allows selection to be completed at a young age. Pedegree records may be used to select animals for performance or progeny testing in multi-stage selection scheme.
d. It allows selection of box sexes for sex-limited traits *i.e.* bulls can be selected on the milk records of their female relatives.

Disadvantages:

a. Accuracy, relative to alternative selection procedures is usually low.

b. Too much emphasis on relatives, especially remote relatives, greatly reduces genetic progress.
c. Progeny of favoured parents are often environmentally favoured.
d. Relatives often make records under quite different environments, thus introducing non random bases into the selection system.

Progeny Testing

In this method we evaluate the breeding value by a study of the expression of the trait in its off springs. Individuality tells us what an animal seems to be, his pedegree tells us what he ought to be, but the performance of his progeny tells us what he is.

Progency testing is, of course, a two-stage selection system because some preliminary selection determines which animals first produce progeny followed by further culling of these which produce poor progeny.

Advantages:

a. High accuracy when many progeny are obtained.

Disadvantages:

a. Long generation interval.
b. Requires high reproductive rate.
c. Low selection intensity.

Show Ring Selection

Selection on the basis of show ring performance has had considerable value in the past. Essentially this selection has been directed towards bringing the conformation of the animal to some ideal conformation. This improvement has been based on two goals: (i) improvement conformation, and (ii) correlated response. Improvement of conformation has economic value because a part of the sale price is determined by the conformation of the individual. The ideal type was chosen so that, in the opinion of the judges, the animal possessing this conformation was most likely to be a profitable producer. In other words, the judges were attempting to stress traits of conformation which are corrected with productive ability.

With the advent of record keeping it was found that direct selection for performance traits resulted in much faster progress than selection through correlated conformation traits. Also, when subjected to intensive study, many of the correlations between performance and show ring were found to be of non-genetic origin. If the correlations are of genetic origin, direct selection for performance should improve conformation as well as the reverse situation. The show ring has been a good forum for discussion of what constitutes ideal type and good management and has produced dramatic changes in the conformation of some species. This has resulted primarily from education of the breeders, however, for most animals which are presented in the ring are good and selection differential among these animals is usually so small as to produce little change.

Advantages:

1. It enables breeders to exchange ideas and experience.
2. It allows comparisons among superior animals both within and between breeds.
3. It allows new breeders to make contact with established breeders.

Disadvantages:

1. Emphasis is usually placed on traits of little economic importance.
2. Clever fitting and showmanship can mask defects of various kinds.
3. Differences between exhibited animals are usually small.
4. Conformation and production traits usually have low genetic correlations.

Choosing Traits for Selection

Many factors enter into the choice of traits to be selected for. The following ones are the most important.

1. The goal of the selection programme
2. The habitability of the traits
3. The economic value of improvement in each trait.
4. The range in variation of each trait.
5. Correlation among the traits.
6. The cost of the selection programme.

Selection Goals

Often the goal of the selection programme makes the choice of traits quite obvious. The breeder of the race horses must select for speed if he is to be successful and his choice of traits are limited to alternative ways to measure speed.

Similarly, the breeder of dairy cattle generally sets out to breed cows with superior milk production cl1aracteristics. Thus, his choice of traits is specified by his selection goals.

Heritability

Heritability is defined to be the fraction of the superiority of parents which is, on the average, transmitted to their off-springs. To explain habitability in simpler words: Heritability tells us how much of the observable differences in the animal is caused by genes and how much by environment. Heritability for the same characteristics may vary from one population to another and also may vary from one characteristic to another even ink the same population.

The ability to recognise the breeding values or transmitting abilities of animals is closely associated with heritability.

If the heritability is high for a trait, we can proceed straight way to adopt a system of mass selection of superior animals, with little attention to pedegree information, collateral relatives, progeny test or inbreeding and genetic improvement in that trait is low, genetic progress may be disappointing with mass selection and greater attention should be paid to pedegree records, family information and use of progeny tests.

Variability of the Trait

Selection operates on the variability in expression of the trait uniform for a trait. there will be little selection response because any selected groups of parents will not be much better than those not selected. Some traits are much more valuable than others. thus the innate variation of the traits should be carefully considered in choosing traits for selection. Variation can be increased by improving exotic types and sometimes this can result in new combination of genes which are superior to either parent type.

Correlated Traits

Sometimes traits tend to be inherited together. These correlations may arise in several ways.

1. The traits may be of different measures of some underlying trait. For example. weight and height are both measures of body size. thus taller animals are usually heavier and these two traits are said to be correlated.
2. If the same genes produce response in several traits. those traits will be correlated. This condition is referred to as pleiotrophy.

Correlated responses are common. Selection for increased milk yield produces a correlated decrease in the per cent of fat in the milk of dairy cows.

Thus. both direct and correlated responses result from selection and some correlated responses are positive while others negative. Correlated response may be advantageously used in selection programme. For example feed efficiency is expensive to measure because it requires both weight gain and feed intake on each individual, whereas weight gain requires neither feed weight nor individual feeding.

In summary. definite goals are essential for a successful selection programme. The success in achieving these goals depends on the existence of genetic differences. the degree to which phenotype differences are heritable and the correlated responses in other traits. In comparing the selection programme, the breeder must evaluate the value of the expected response and the cost of the programme relative to the costs and responses of alternative selection programmes.

GENETIC IMPROVEMENT OF DAIRY CATTLE

Dairymen are constantly concerned with improvement of their herds. Dairy cattle improvement is expressed most frequently as increased milk production and less often in terms of changes in body conformation or "type." Improvement in milk production can be considered as being environmental or genetic.

Both environmental and genetic factors contribute to variation in milk production. There are environmental and genetic limitations of production. More importantly, management of environment and genetic factors can result in improved performance, *i.e.,* greater milk production.

Genetic Principles of Repeatability

Understanding the concept and knowing the repeatability of economically important traits help considerably in deciding which cows to cull. Repeatability also is important in sire proofs. All repeatability estimates represent some type of correlation. The trait one is considering, and the times and places it was measured, influence the estimate. The correlation between 305-day mature equivalent milk records of first-calf heifers, and their records as second-calf heifers, is about 0.5.

So milk records are said to have a repeatability of 0.5. Conception rate from year-to-year with the same cow has a repeatability of about zero, indicating that there is little or no relationship (correlation) between the number of services it takes for a cow to become pregnant each time. These estimates are useful in making culling and management decisions, since they help us make the best possible guess at how well the cow will do next year if we know how well she did this year. For milk yield, 50 per cent of the apparent superiority (or inferiority) of the cow for this year will appear next year on the average.

If a cow gave 1000 pounds more than her herdmates this year, the best prediction of her yield next year will be 500 pounds more. Naturally, what she actually does is another story; she could be much better or much worse than our guess. The same process is used for below average cows; only one-half of their apparent inferiority will appear the following year on the average. Repeatability is included in the calculation of estimated relative producing ability in DHI. Only a fraction of the cow's performance is permanent (repeatable). Superior cows may do even better next year, but on the average they do worse; inferior cows on the average do better.

Genetic Principle of Heritability

Heritability is a measure of the degree to which a trait such as milk yield is genetically determined. Although not all of the genetic contribution is included, essentially all which is available to a breeder who is trying to select within a breed is included. Obviously heritability is an important factor among the several factors determining how much genetic improvement can be made in any characteristic.

One of the disadvantages with the dairy cow, in attempts to make genetic progress, is that most of the economically important traits can be measured only in the female. Another disadvantage is that heritabilities are low or close to zero in several cases. A zero heritability would mean that no genetic change could be expected from usual selection techniques.

A practical idea of the meaning of heritability can be obtained easily. A genetically average group of males is mated to a large group of high producing first calf heifers. These heifers average 1000 pounds of milk more than their contemporaries. Assume also that no extraordinary changes in environment have occurred by the time that daughters from these matings freshen and complete their first record.

How much milk will the daughters produce compared to their herdmates? Answer: about 100 to 150 pounds more. Reason: milk yield has a heritability of about 20 to 30 per cent, and since the sires used were genetically average, and sires and dams contribute about equally to the genotype of their offspring, only 10 to 15 per cent of the superiority of the dams will be seen in the offspring.

A good practical definition of heritability is, therefore, the fraction of the merit (or lack of merit) observed in the parents which can be seen in the offspring, with merit being defined as a deviation from the population average. Note that several reproductive traits have very low heritabilities. Milk yield would be said to be moderately heritable; fat and protein percentages are highly heritable.

One possibility exists for genetic improvement of important traits with zero or very low heritability. This revolves around what geneticists call non-additive genetic variance. If there is appreciable non-additive genetic variance, it can be taken advantage of by crossbreeding. Heterosis (hybrid vigour) is a reflection of this type of genetic variability. Unfortunately, although there is some heterosis in reproductive performance, longevity, livability and milk yield, it is not a large amount.

Crossbreeding has several disadvantages for dairy cattle, although this discussion cannot be covered here. Few people are ready to advise taking advantage of non-additive genetic variance in dairy cattle at the present time, and few cattleman are attempting to do so.

Reproductive efficiency can be improved considerably by improving management practices while selection is being practiced for more highly heritable traits.

SYSTEMS OF BREEDING

The ultimate aim of the breeder is to evolve outstanding and improved type of animals which can render better service to man. Selection and system of breeding constitute the only tools available to the breeder for improvement of animals. Since new genes can not be created though they can be recombined into more desirable groupings.

Systems of breeding has been broadly divided as under:

1. In breeding -breeding of the related animals.
2. Out breeding -breeding of the unrelated animals.

Inbreeding

Inbreeding is a mating system in which individuals mated are more closely related than the average of the population from which they come.

It means the mating of males and females which are related. Animals deemed to be related only when they have one or more ancestors in common on the first 4-6 generations of their pedegree. The intensity of inbreeding depends upon the degree of relationship. Close inbreeding denotes mating of closely related individuals like dam to son (mother x son) or sire to daughter (father x daughter) or full brothers to full sisters.

In breeding makes more pairs of genes in the population homozygous. Wherever there is inbreeding. there will be one or more common ancestors from which, part of the gene samples (gametes) have arisen.

Close Breeding

This means the mating of full sister to full brother or sire to his daughter or dam to her son. These types of matings should be only when both parents are outstanding individuals, and then only at the increased risk of bringing undesirable recessive genes into homozygous form in the progeny.

Advantages:

1. Undesirable recessive genes may be discovered and eliminated by further testing in this line.
2. The progeny are more uniform than out-bred progeny.

Disadvantage:

1. The undesirable characteristics are intensified in the progeny if unfavourable gene segregation occurs.
2. It has been observed that the progeny become more susceptible to diseases.
3. Breeding problems and reproductive failure generally increase..
4. It is difficult to find out the stage of breeding at which it should be discontinued in order to avoid the bad effects of the system.

Line ZBreeding

This means the mating of animals of wider degrees of relationship than those selected for close breeding. It promotes uniformity in the character. Homozygosity is not reached so quickly as in close breeding. Neither desirable nor harmful characters are developed so quickly. It is a slowed method for the fixation of hereditary outstanding bull or cow and the progeny is mentioned as being line-bred to certain ancestors.

Advantages:

1. Increased uniformity.
2. The dangers involved in close breeding can be reduced.

Disadvantages:

1. The chief danger in live breeding is that the breeder will select the animal for pedegree giving no consideration to real individual merit. This may in some cases result in a few generations which receive no benefits from selection.

Consequences of in-breeding

The effects of both close breeding and line breeding are similar. The only difference is in the degree of their intensity. They are more intense in close breeding and less so in line breeding.

These may be described as follows:

1. It increases homozygosity (like alleles) and decreases heterozygosity (dissimilar genes) and hence favours the development of genetic uniformity amongst the animals.
2. It is the best method of getting true strains from unknown stocks as it sorts out the characters in the homozygous condition and thus help in the selection of undesirable individuals and culling of the undesirable individuals. Strains which breed true are not obtained in animals. however, and in plants are obtained only when close inbreeding is accompanied by intense selection.
3. The outward effects of in-breeding may include the following

Effect on growth rate: With the exception of a few laboratory animals, some moderate decrease in growth rate and mature weight has been noted. The studies on rats on the other hand have shown that they withstand inbreeding (up to 100 generations over 40 years) without any apparent bad effect on growth rate. The reproductive rate of rats is very high, however, relative to farm stock.

Effect on reproductive performance: In almost all the cases a reduction in the reproductive efficiency has been noted. It may delay testicular development and puberty, reduce gametogenesis or increase the embryonic death rate.

Effect on vigour: It has been noted that the death rates among the in-bred groups of animals are higher than those in out-bred ones. They are also adversely affected by environmental conditions. They have higher percentage of Runs (animals which are undesirable due to various reasons).

Effect on production: Production traits usually show moderate decrease with increased inbreeding.

Appearance of abnormalities: Hereditary abnormalities or lethal factors are likely to appear more often in the inbred animals than in the out bred ones.

Of the two components of inbreeding *i.e.* close breeding and live breeding, it is safer to go for the later one. A few recommendations have also been made With regard to them.

Inbreeding is useful if we want to evolve a more uniform line in any animal species. A line is often desirable for scientific purpose when animals or plants for a number of lines can be mated to get a breed with many desirable qualities. However, that is the job of a scientist or precisely a geneticist. A fine line of animals may not be commercially profitable at all. A farmer is, therefore, advised not to conduct this experiment inbreeding unless he knows genetics.

Out breeding: Out breeding is the breeding of unrelated animals and this involves the following types of breeding.

1. Out crossing
2. Cross-breeding
3. Hybridization
4. Grading up.

The genetic effects of out-breeding are opposite to those of inbreeding. Whereas inbreeding increases homozygosity, out breeding tends to make more pairs of genes heterozygous. Therefore, out-bred animals are less likely to breed true than are in-bred animals.

Out-crossing: Mating of unselected animals within the same pure breed is termed out- crossing. The animals mated have no common ancestors on either side of their pedegree up to 4 -6 generations and the off-spring of such a mating is known as the outcross. Traits which are largely additively genetic and which have a high heritability, respond quickly to out crossing. In a period of many years, the various breeds in many western countries registered progress mainly though a system of out-crossing in which the best available, but unrelated sires were continuously selected for use on the females in a herd or flock.

Advantages:

1. This methods is highly effective for characters that are largely under the control of genes with additive effects *i.e.* milk production, growth rate in beef cattle, etc.

2. It is an effective system for genetic improvement if carefully combined with selection.
3. It is the best method for most herds.

Species Hybridization

Crossing of individuals from two species is called hybridization. This is the widest possible out breeding system. It is possible only between related species which might have descended from common parent stock somewhere back in the evolutionary process. In the course of time, evolutionary forces might have set them off into distinct species. A well known example for species hybridization is the mule which is a cross between the jack ass and the mare. The mule is valued for its ability to work hard under the most adverse conditions. The mules are generally sterile.

Similarly, European cattle and American bison when crossed produce sterile males and fertile females. By back-crossing the females to bison and cattle attempts are being made to form a new breed of cattle-the cattalo. Species crosses are possible only when two species are related closely enough. These crosses usually result in much heterosis (a phenomenon in which the crosses of unrelated individuals often result in progeny with increased vigour (much above their parents). However, the progeny of species crosses have very low or nil fertility.

Grading is a process by which a few purebred sires can rather quickly transform a non descriptive population into a group of pure breeds. The grading process does not create anything new but it may transfer the good qualities of an improved breed. For a grading up programme, it is advisable to use a breed that I has thrived well under local condition. Otherwise graded animals may not adapt themselves to the local environment.

Advantages:

1. Pure breeds can be obtained just after a few generation (after 7th to 8th generations).
2. The start can be made with a little money in comparison to the purchase of an entire herd of purebreds.
3. It helps to prove the potentialities of the sire and adds to its market value.
4. It is a good start for new breeders who can slowly change over to pure breed systems.

Limitations of grading:

1. Pure breeds are not always better than grade or country animals for the use to be made of them.
2. Pure breed stocks which give good results in one set of environmental conditions do not always give favourable results in some different

environmental set up. The pure breed dairy cattle from temperate zones often degenerate when used in tropical areas. Moreover, their off springs fail to show vigour and constitution for high reproduction. To make grading successful the pure breed must have the ability to perfonn under the environmental set up where their offspring is going to perform.

Improvement of Dairy Cattle in India

The dairy fanning business in many tropical and subtropical countries are characterized by large number of cattle and low yield of milk. For example, India possesses over 230 million heads of cattle and buffaloes, which comes to more than one-fifth of the total world population of cattle and buffaloes. The average annual production of an Indian cow is only 173 kg as against an average yield per milking cow of 3,710 kg in Denmark, 3,250 kg in Switzerland and 3280 kg in the U.S.A.

Twenty six breeds of cattle and seven breeds of buffaloes have been recognised at present in India. However, only a small proportion of the cattle and buffaloes belong to these purebred breeds. The vast majority (more than 75 per cent) are intermixed and do not belong to any specific breed. They are classed as nondescript.

Cattle remain the draught animals in many of these regions. As a result selection over the past many centuries has been to meet the requirements of draught and agriculture. Selection pressure was applied for better capacity and fitness. This has resulted in producing excellent draught animals. Unfortunately, milk production and draught capacity are traits which are genetically negatively correlated. When milk yield capacity goes up, draught capacity goes down and vice versa. As a result attempts to develop and improve some of the dual purpose breeds which are useful for both milk and work, could not progress very far. We could have only a compromise in dual purpose breeds with medium milk production and medium work capacity. If further genetic improvement of either of the two traits is attempted, a set back with respect to the other should be expected.

In India, at present 94.3 per cent of the milch cows yield less than one kg. milk per day. Only 0.5 per cent yield more than 2 kg per day. Buffaloes are better milk producers. About 18 -20 per cent of the milch buffaloes yield more than 2 kg per day and only 19 per cent yield less than one kg per day. The reasons for such a low level of production are listed below.

1. Acute shortage of feed and fodders.
2. Excess cattle numbers.
3. Poor genetic potential for milk production.

4. The smallness of land holdings and consequently small dairy units make them economically unviable. It also acts against introduction of advanced techniques.
5. Adverse climatic conditions of the tropics.
6. Poor grazing and environmental factors.
7. Inadequate marketing facilities for the products.

The diversity of the breeding stock and the variation available in economic traits of cattle and buffaloes in the country offer greater challenge and scope for their improvement for the animal breeder. At the same time, the task of improving the genetic make up of a large number of extremely diverse, non-descript low producers is a colossal one. Anyone system of breeding can not be applied uniformly to all the animals in all the areas.

2

Animal Breeding and Genetic Theory

The development of selection index theory 70–80 years ago formally embedded economic principles in the practice of animal breeding, and many of our leading practitioners have made important contributions in this area. Animal breeders have relatively infrequently drawn on economic theory and instead have developed their own simple and practical tool kit to address the issues of prioritizing traits for genetic selection decisions and when undertaking cost benefit analysis. We know from our quantitative work that models do not have to be true in order to be useful, and the tool kit we have has certainly proved useful. It is also much more accessible than many of the highly complex theories that exist in economics.

In this editorial, I argue that principles of economics can help us to better understand successes and failures in animal breeding. Notable successes include the massive and sustained improvements in the productivity of intensively farmed livestock species, particularly pigs and poultry systems but also dairy production systems. Failures include attempts to move exotic genotypes into unsuitable environments, and the relatively low return to substantial investment in ongoing genetic improvement in extensive and low-input production systems in both the developed and developing worlds.

When compared with most farm management alternatives, genetic improvement is an extremely slow and quite clumsy technology option. For many of our successes, these deficiencies have been offset by the permanent and cumulative nature of genetic improvement, and the ability to remould farm management systems and practices to suit the new genotypes that have evolved.

The main farmed species and farming systems are highly varied in terms of how genetic improvement can be achieved and distributed, and undoubtedly, these differences have contributed to the successes and failures. However, economic principles of market failure, vertical integration, value capture, intellectual property (IP) rights, efficiency because of competition and disruptive technologies interact with these differences and so are also key contributors to success. A relatively simple understanding of these factors can be useful in identifying potential remedies for our not infrequent failures. In the absence of this understanding, we as animal breeders may be tempted to continue serving up the same old approaches using ever more complex statistical methods and highly expensive genomic tests that fail to free industries from the shackles of low adoption and ineffective distribution of genetic improvement.

Attempts to address market failure in animal breeding have a long history, most notably in the establishment of novel and distinctive physical characteristics (horns, ears, coat colour, body shape, etc.) which effectively brand breeds and sometimes lines within breeds such that they can achieve a market premium. Differentiation relative to unimproved commercial or cross-bred stocks is critical, but differentiation relative to lower-performing competing breeds and lines within breeds is also very important.

Unfortunately, there is also a long history of selection focus drifting almost fully onto distinguishing characteristics, and a consequent loss of focus on the production characteristics required for ongoing improvement. Within breeds, failure of commercial farmers to be able to physically observe superior genetic performance, either at all or within a reasonable time frame, has been and still is a major impediment to improvement in many extensive livestock farming systems. It seems no coincidence that the least successful livestock species for within-breed improvement are commonly those with the most extensive farming systems.

Vertical integration, whereby a single commercial entity maintains ownership of animals from breeding through to processing, and even marketing and distribution in the livestock value chain, often overcomes many problems of market failure. Improved animal identification, measurement technologies and information systems are also helping to create more transparent markets. This is important, because vertical integration usually requires large scale to justify a breeding programme. Increasing scale also increases risk. Vertical integration is often but not always a key contributor to the success of intensive livestock breeding programmes. Unfortunately, extensive livestock industries are already subject to variation in the production environment. Globalization of markets appears to be exacerbating fluctuations in input and output prices. It seems therefore that vertical integration is unlikely to accelerate rapidly in extensive livestock industries, because of the concentration and magnification of risks.

Value capture is a critical contributor to the success of many plant breeding programmes. For example, the inability of farmers to multiply their own seed from hybrid corn creates a captive market for seed, which provides revenue for ongoing investment in genetic improvement. Value capture is also a key and often highly controversial factor in animal breeding programmes. The massive and rapid international dissemination of elite Holstein dairy cattle semen leaves AI companies on a desperate treadmill striving to maintain a point of difference, while their competitors' opportunities to identify elite sons of the top sires globally have historically been limited only by the relative sizes and efficiency of their respective progeny testing schemes. But still, a historically strong emphasis towards the need for high reliability, itself a conservative measure of risk compared with accuracy, has helped dairy AI companies to stave off potential competition from semen from high-merit unproven young bulls.

The trend in animal breeding programmes over many decades has been towards privatization of breeding programmes, and a switch in publicly funded research effort away from practical breeding and towards a search for new step change and/or disruptive technologies. IP rights pertaining to germplasm are weakly controlled in animal breeding. This motivates prevention of leakage of elite germplasm to competitors, something that has been achieved to a reasonably high extent in pigs, poultry and for some species in aquaculture. This protection seems justified in the context of the developed world, relying on private investment to drive high levels of genetic progress.

In contrast, for developing countries, much of the genetic progress that arises comes about through lateral shifts and exploitation of high-performing germplasm. Indeed, for much of global dairy production, and in extensive livestock industries, genetic progress also comes about through the same means. These lateral shifts occur at the levels of sire sharing within group breeding schemes up to wide-scale international trade in elite germplasm across many species and farming industries. Thus, competition has led to more efficient breeding programmes in some industries, but sharing of germplasm is a substantial contributor to a large amount of realized genetic gain in others. Molecular biologists have long-threatened animal breeders with new approaches that would act as disruptive technologies for the process of achieving animal genetic improvement. After many years of promises, genomic selection appears to be on the verge of offering a major disruption to the global business of dairy cattle breeding. Barriers to market entry by new progeny testing initiatives have revolved around the cost of setting up a scheme sufficient to meet minimum reliability criteria and the time required to achieve results in industry, in an already highly competitive business. These barriers can potentially be broken down by public investment and availability of accurate predictions of the genetic merit of young bulls. It will be interesting to see how the future unfolds. Genomics may yet help reverse the trends of disappointing rates of genetic improvement in extensive species and in developing countries.

Big decisions and consequent changes to approaches are likely for animal breeding over the next decade as food demand grows and new technologies continue to unfold and be exploited. Economics, and an understanding of the past, will help us to shape the future to take full advantage of the huge potential of livestock genetic improvement.

ENDANGERED WILD ANIMALS

The principal threat to wild animals arises from human population growth and the aspirations of the human population for an improved quality of life. The human population, now (2006) 6.5 billion, increased exponentially from an historic base of approximately 300 million at the time of Christ to about 800 million by 1800, and increased fourfold during the twentieth century alone. It is expected to reach 9 billion by 2050. The growth in human industry (economic activity) is even more revealing than the increase in population. During the twentieth century, a fourfold increase in the population was accompanied by an 18-fold increase in the economic output. During that time, we lost about 40% of the planet's forests and 10% of the coral reefs (both major repositories of biodiversity) with additional threats to mangrove swamps, wetlands and savannah.

Extinction is of course a normal part of the life process; for example, since birds first appeared 130 million years ago, 500 000 species are said to have existed, the maximum number extant at any one time being 11 500. Today there are 9946 species of birds, about 1200 of which are threatened (listed as critical, endangered or vulnerable by the Species Survival Commission Red Data List).

One hundred species have become extinct during the last 100 years. However, these statistics hide the fact that the rate of extinction, and the level of threat, is about 100-fold higher today than it has been historically, for all groups of vertebrates. For mammals, 1100 of 4763 species are threatened; for reptiles, amphibians and fishes the situation is less well known and the proportions threatened are underestimated (namely 253 out of 7970 species, 124 out of 4950 and 734 out of 25 000, respectively). The evaluation of the extinction threat to a species is an imprecise science, but has been put on a more quantitative basis by the development of criteria for endangerment by Mace & Lande (1991). The maintenance of genetic diversity is one of the goals (G3)a of precision animal breeding.

Ex Situ Versus in Situ Breeding

In making breeding decisions for endangered species it is clearly advantageous to ensure the survival of animals in their home range. However, where this is not possible it is better to remove individuals to an alternative, safer, environment than to allow them to be hunted to extinction. Alternative locations include zoos or wildlife parks, and may be found in a country far from the animal's original habitat. This potentially gives rise to conflicts of interest. Removing individual animals from their home range reduces the number *in situ* thereby putting further pressure on those remaining. Where the *ex situ* site is far from the region or country of origin, there may be political pressures against removal. Local extinction following removal may result in the original habitat being irrevocably lost as a result of the lack of pressure to maintain the population locally, making reintroduction difficult. Clearly where possible it is preferable to achieve a combination of *in situ* and *ex situ* approaches.

Conservation Organizations and Structures

There are two broad categories of organization involved in the management of populations of wild animals: those based in zoos, aquaria and wildlife parks, and those overseen by the United Nations. In general, those representing the UN oversee information on a global scale, collecting and managing information with a view to developing priorities for habitat and species conservation both *in situ* and *ex situ*. Zoos, aquaria and wildlife parks principally oversee conservation breeding and management *ex situ*, but are also becoming increasingly involved in *in situ* conservation. The tools used centre on stud books, which is where the pedigree and other information is recorded, and software packages to manage the stud books.

Four United Nations bodies are involved in different aspects of conservation of biodiversity: the World Conservation Union (WCU; formally the International Union for the Conservation of Nature and Natural Resources, IUCN); the United Nations Environment Programme (UNEP); the United Nations Development Programme; and the Food and Agriculture Organization (FAO). The WCU (IUCN) established Commissions on Ecosystems Management, Education and Communication, Environmental, Economic and Social Policy, Environmental Law, the World Commission on Protected Areas and the Species Survival Commission. The last of these, the SSC, is the largest, with 120 specialist groups and task forces to identify threats to groups of taxa and recommend conservation priorities and actions. These specialist groups include the Conservation Breeding Specialist Group (CBSG) which, working through Conservation Assessment and Management Plans and Global Captive Action Plans, identifies species in need of propagation in captivity with a view to subsequent release. The principal role of the CBSG is to facilitate decision making by arranging meetings of appropriate experts. By convening and hosting meetings of interested parties, this body develops Habitat Management Plans for individual species (population and habitat viability assessments, PHVA) or groups and geographical regions (Conservation Action Management Plans, CAMPs). PHVAs use software developed for the purpose, which applies population dynamics techniques, and takes into account demographic events at an individual animal level. The CAMPs use databases to formulate recommendations on actions required to manage populations or habitats. The level applied is wider than in the case of PHVAs, as CAMPs deal with broad taxonomic groups or geographical regions. Although their objectives appear to overlap, the structures organized by the zoo associations run in parallel with CBSG, through common membership; the same personnel are frequently involved in both sets of bodies.

Through UNEP, the World Commission on Environment and Development (the Bruntland Commission) in 1987 highlighted the importance of biodiversity and sustainability, initiating a process of international discussion culminating in the United Nations Conference on Environment and Development in Rio de Janeiro in 1992. This conference led to the Convention on Biological Diversity, which has now been ratified in 188 countries. The three objectives of the Convention are the conservation of biodiversity, the sustainable use of its components and the equitable sharing between societies of benefits arising from the exploitation of genetic resources. In 2002, the parties to the Convention adopted a strategic plan, focusing on reducing the rate of loss of biodiversity globally, regionally and locally by 2010. The seven strategies adopted as means to achieving the objectives of the Convention include: (i) 'Reducing the rate of loss of the components of biodiversity, including: (a) biomes, habitats and ecosystems, (b) species and populations, and (c) genetic diversity' and (ii) 'Promoting sustainable use of biodiversity'.

Notable advances have been made during the past 20 years in the management of endangered animals in captive populations held by zoos and wildlife parks. Stud books are held by individuals, who are most frequently animal managers and keepers in zoos. Permission to hold a stud book is agreed with the local zoo association (the European Association of Zoos and Aquaria (EAZA) in Europe or the American Zoo and Aquarium Association (AZA) in the USA), and ratified by the AZA Wildlife Conservation and Management Committee (WCMC) and the World Association of Zoos and Aquaria (WAZA). Stud books are managed using an internationally accepted software package, which includes compatible software for keeping animal records and medical records. In North America, the AZA represents more than 200 zoos and operates for conservation breeding purposes through the

WCMC. In turn the WCMC organizes Taxon Advisory Groups (TAGs, of which there are 46), which are composed of experts on particular groups of animals (taxa), recommending breeding actions to AZA institutions, evaluating the need for captive breeding and assessing the space available in zoos. Their members have expertise in taxonomy, assisted breeding, contraception and wild populations and habitats, as well as educational and training programmes. In North America TAGs develop two kinds of breeding plan, Species Survival Plans (SSPs) and Population Management Plans (PMPs). SSPs (of which there are currently 107, representing 161 species) include genetic and demographic analyses of captive populations, and make individual breeding recommendations for both *in situ* and *ex situ* populations. PMPs (of which there are 282) fulfil the same function as SSPs, but for the *ex situ* population only. Both SSPs and PMPs work through stud books (of which there are more than 400 in North America). TAGs also develop action plans with priorities for *in situ* conservation requirements. The SSP programme started in 1981, and PMPs in 1994.

A similar structure operates outside North America. In Europe, 200 zoos in 25 countries are involved in breeding programmes. The equivalent of SSPs and PMPs are European Endangered Species Programmes (EEPs, of which there are 151) and European Stud Books (ESBs; 140), and these are run through Taxon Advisory Groups (n=40) reporting to the EAZA. Similar structures are in place in Africa, Australasia and Asia.

Conservation strategies for zoos worldwide are brought together by WAZA. The 2005 second edition of *Building a future for wildlife: the WAZA conservation strategy*, published by WAZA, provides a blueprint for the roles zoos and aquaria play in conservation of wildlife and their ecosystems. It represents a common philosophy for zoos and aquariums worldwide and defines the policies required to achieve their conservation goals.

Conservation Breeding and Release

There have been notable successes for conservation breeding, which have shown that with enthusiasm, commitment and an ability to collaborate, it is possible to ensure the survival of species that would otherwise have been lost. A good example is that of the Arabian oryx (*Oryx leucoryx*), a gazelle hunted to extinction in its natural range (the southern Arabian peninsula, Aden, Yemen and Oman) in 1972. A survival plan was put in place, largely initiated by the Fauna Preservation Society of the UK (now Fauna and Flora International), between 1961 and 1982 which, with 3 animals wild-caught in Aden in 1962 and 16 others from zoos taken to Pheonix, Arizona, bred a herd of 35 by 1972 and 106 by 1977. Offspring were returned to the Middle East in 1978 and to their original range in Oman in 1982. The cultural benefit of this reintroduction programme to the inhabitants of that region is hard to overestimate, and their involvement in the animals' care is evidence of the important place this oryx occupies in the culture of the desert-living nomads of that area.

Other successes of conservation breeding have been the scimitar-horned oryx, Pére David's deer, Przewalski's horse, the black-footed ferret, the Mauritius kestrel, the Puerto Rican parrot, the black and white ruffed lemur, the Egyptian tortoise, the Partula snail and the Wartbiter cricket. In each case the principles of precision animal breeding have been applied, because the animals involved were bred for a purpose and with limitation of loss of genetic variation (G3) in mind. In some cases the pedigrees of the animals used were not well known, but nonetheless, the principles were applied.

In the case of animals bred for subsequent release from captivity, there is a need for the individual animals involved to be isolated from domesticating influences (Woodworth *et al.* 2002), and to be trained in behaviours required for an unsupported existence. Will an animal raised in captivity be able to hunt, or to identify predators or poisonous plants? Examples of species in which these processes are important include most notably primates, for instance the golden lion tamarin, which has been released into its former range in Brazil.

There is also a need to carefully consider the genetic composition of the released population and its impact on the animals remaining in the captive group. Maximizing genetic diversity in the released animals may deleteriously affect that of the captive group, which should be avoided if further releases are planned, and the captive population is to be maintained. Releasing individuals over-represented by progeny benefits the captive subpopulation, because in a captive breeding programme these animals would not be bred, and their reintroduction will free spaces for further growth of the captive population. Where there is uncertainty over the survival of reintroduced animals this strategy is appropriate, because it protects the genetic diversity of the remaining captive animals. On the other hand this strategy may not maximize the survival chances of the reintroduced group, because that benefits from maximizing genetic diversity in the reintroduced animals. These trade-offs have been modelled for four captive-bred species with different breeding histories by Earnhardt (1999).

Tissue Banking and Conservation by Nuclear Transfer

Gamete storage techniques (sperm and oocyte freezing; germplasm preservation) have given rise to the possibility of the 'frozen zoo'. Recently this has been expanded by the opportunity to propagate nuclear DNA through somatic nuclear transfer (e.g. Gómez *et al.* 2004), offering the possibility to store tissue in the form of cell cultures, which can be frozen, for subsequent nuclear transfer to an enucleated oocyte, through which cloning can benefit diversity (Woolliams & Wilmut 1999). A number of issues arise with these techniques. Firstly, it is relatively easy to freeze gametes, but more difficult to ensure they are fertile on thawing. Secondly, the numbers of endangered animals to which gametes can be transferred are limited, and this limits the information available on processes such as superovulation and induction of oestrus, which are required as part of the technique. Thirdly, somatic cells in culture may undergo chromosomal reorganizations rendering them unfit for nuclear transfer. In somatic nuclear transfer, mitochondria are transferred from the donor somatic cell to the recipient oocyte, and the offspring may be mitochondrial chimaeras; this is particularly likely if a surrogate oocyte (from a related but non-endangered species) is used as recipient. Lastly, which individual animals to preserve? There is an argument to conserve the broadest set of alleles possible from a population; appropriate animals can be identified by genotyping, by measuring a set of quantitative traits or on the basis of coefficients of relationship (Lamberson *et al.* 2002). However, a random stratified sample of the population representing the gene flows in the population as identified by pedigree, performance and molecular analysis may be least at risk from deleterious genes.

The storage in frozen form of genetic material from animals is therefore much more problematic than for plants. But despite the difficulties, it does offer great opportunities, and several centres are engaged in establishing such tissue banks (including in the UK the Institute of Zoology, London and the University of Nottingham).

Opportunities for this Century

One of the key questions arising in making conservation decisions is the identity of the *ESU*, the 'population unit meriting special management' (Ryder 1986). This is defined as 'a set of populations that is morphologically and genetically distinct from other similar populations or a set of populations with a distinct evolutionary history'. This may be smaller than the species, and may not necessarily represent a reproductively isolated population: it may be based on genetic, phenotypic or behavioural criteria. The US Endangered Species Act recognizes ESUs as the unit requiring protection. The identification of the ESU for a particular species most frequently depends upon molecular genetic information, such as mitochondrial DNA and microsatellite polymorphisms. Thus, the application of molecular genetics techniques to conservation questions is of paramount importance. Examples of ESUs identified in this way are the dusky seaside sparrow and the red wolf (although in each of these cases the evidence has been challenged). An example of a species found not to represent an ESU is the Cape Verde kite (Johnson *et al.* 2005) though this study did identify two other species of kite as phylogenetically distinct, the yellow-billed kites from South Africa and Madagascar and those of northern Africa. More needs to be done to define how ESUs are determined for each species, and to apply these techniques more widely.

The developments in requirement R3 that will arise during this century will help in the recognition of selection pressures that lead to adaptation to captivity for conserved wild species. The availability of DNA information will play an increasing role in the management of diversity within an ESU (after having defined it); however it is important to recognize that the scope for managing diversity beyond best practice using pedigrees is limited. Goal G3 of precision animal breeding can be considerably advanced by more informed rural planning to avoid wild populations being fragmented into small isolated groups.

In coping with climate change and the increasing need for conservation work that will arise from it if present predictions are even close to reality, there will be a need for committed personnel, highly productive systems and structures, and well-defined objectives. Above all there will be a need for more 'spaces' for endangered animals in conservation breeding programmes.

Companion Animals, Pets and Animals for Recreation

Large numbers of animals are bred as pets, most frequently by professional breeders. The number bred by individuals for their own use is small by comparison. The range of species bred for this purpose is widening continually, and it is now possible to buy as pets, wallabies, miniature horses and donkeys, and a wide range of other vertebrates. Most pets, however, are dogs or cats; the size of these populations is an indication of their potential to cause mischief: there are 6.7 million dogs and 7.5 million domestic cats in the UK. There is also a feral cat population of 800 000, and each year about 135 000 dogs stray, of which about 10% are put down. There are approximately 975 000 equids (not exclusively horses). For each of these species there are examples of successes and failures in the application of the principles of precision animal breeding, sometimes with highly unfavourable genetic consequences. Online Mendelian Inheritance in Animals lists more inherited disorders in dogs than any other species excluding humans and mice (479, of which only 47 have been characterized at a molecular level), with cattle second (366, 31 characterized) and cats third (273, 13 characterized). In many cases these are diseases with human homologues. Nonetheless, significant advances are now being made in the recognition of problems associated with certain breeds, and in structures which should lead to more widespread use of appropriate techniques.

Cats and Dogs

Selection in the breeding of cats and dogs is based on satisfying human aesthetic criteria rather than those ensuring fitness in a wild environment. As a result deleterious mutations have become widespread in populations, through inbreeding used to 'fix' desirable traits).

In the UK the breeding of dogs is controlled by the Breeding of Dogs Acts of 1973 and 1992, and the Breeding and Sale of Dogs (Welfare) Act 1999. This legislation licences puppy farms and ensures that the dogs are suitably accommodated, fed, exercised and protected from disease and fire. The 1999 Act provides that bitches are not mated until they are 1-year-old, that they have no more than one litter per year and that they give birth to no more than six litters in a lifetime. The Act also requires that accurate breeding records are maintained. This legislation, together with a range of other legislation relating to the welfare of farmed and non-farmed animals, is being brought together under a new bill (the Animal Welfare Bill) presently before the UK parliament. In addition to the above legislation providing for breeding of puppies, breeding of certain types of dog for fighting is prohibited under the Dangerous Dogs Act 1991, which 'prohibits persons from having in their possession or custody dogs belonging to types bred for fighting; imposes restrictions in respect of such dogs pending the coming into force of the prohibition; enables restrictions to be imposed in relation to other types of dog which present a serious danger to the public, and makes further provision for securing that dogs are kept under proper control'.

None of this legislation covers selection in breeding, however, and as described above, many breeds now suffer from significant levels of inherited disease. The ways in which Kennel Clubs and dog shows have operated in the past have led to their rapid dissemination. Therefore, for example: (i) a number of breeds have been established with too few founders, (ii) stud books are closed for some breeds, preventing introgression of new genetic material and leading to inbreeding, (iii) selection has often been based on inappropriate type criteria, and (iv) even if the founders were sufficiently diverse genetically, clubs have frequently failed to exert adequate management of the genetic diversity generation by generation.

Kennel Clubs are generally aware of these problems. Stud books are becoming 'less closed' and phenotypic screening tests are becoming available and in use for many common inherited disorders. However, these do not identify individuals carrying recessive genes or late onset disorders which can only be diagnosed after the animal has bred. Of greater potential value are molecular markers for disease, which are currently being developed for a wide range of conditions. The development of these DNA-based tests has been facilitated by sequencing of the dog and cat genomes.

These tests will identify carriers of recessive diseases such as copper toxicosis and progressive retinal atrophy, which affects several breeds, as well as mutations responsible for von Willebrand's disease in Shetland sheepdogs, Scottish terriers and Doberman Pinschers (where different mutations are responsible for the condition), phosphofructokinase deficiency syndrome in English Springer and American Cocker spaniels and pyruvate kinase deficiency in Basenjis. These tests require small blood samples and can be carried out on prepubertal animals. DNA tests for progressive retinal atrophy in Irish Setters, copper toxicosis in Bedlington terriers, and hereditary cataract and L-2-hydroxyglutaric aciduria in Staffordshire Bull terriers are already available in the UK through the Animal Health Trust and the Kennel Club. In order to speed eradication of progressive retinal atrophy in Irish Setters, there is a ban on breeding and showing carriers for the condition, which can be identified using the test. This is an example of what is achievable in this field, and it is to be hoped that other tests will be introduced with similar openness of information.

In some countries, these tests are required by law. The Netherlands and Germany have legislation banning breeding of animals with severe defects or breeding that could result in offspring with inherited disease. This is a strong implementation of goal G2. In Sweden it has been proposed to limit the number of litters per parent based on the population size for each breed, which will limit the impact of deleterious genes on the population.

There are nonetheless notable instances of excellence in breeding of dogs in the UK. An example of an organization taking a highly professional approach to breeding working companion animals is the Guide Dogs for the Blind Association. This charity breeds over 1000 guide dog puppies annually, and supports about 5000 guide dogs in use in the UK. About £10 per day is invested in the breeding, training and care of a guide dog, which is a measure of the level of investment that should be applied in breeding working dogs.

In contrast to the controls on dog breeding, the breeding of cats is uncontrolled, presumably because cats live more independent lives. Molecular approaches are less well developed than for dogs, but the mapping of the feline genome has recently been reported by Murphy *et al.* (2000), and DNA tests are being introduced (for instance the scheme for polycystic kidney disease run by the Animal Health Trust and the Langford Feline Diagnostic Service).

Horses

Too little attention has been paid to selection in breeding sport horses. It has been suggested for some time (Cunningham 1989) that speed of racing in thoroughbreds has not improved for some years, possibly through lack of genetic variation, or from attainment of a physiological 'ceiling'. More positively, a recent study (Mota *et al.* 2002) has indicated a small genetic trend in racing time improvement, suggesting that failure to observe any phenotypic improvement may be due to only weak selection pressure since selection has been largely on prize money rather than race times. However, such trends need to be better established.

The Jockey Club rules governing thoroughbred breeding preclude artificial insemination or other assisted reproduction techniques, and this limits genetic improvement but reduces the potential to erode levels of genetic diversity. However, these rules do not apply to other breeds, and there is in particular much to be gained from a systematic approach to breeding horses and ponies for sports such as eventing and polo. This has traditionally been carried out with little or no consideration of performance testing, and as a result opportunities have been missed in international sports.

Opportunities are now emerging, however, through legislation affecting the management of horses. Following the publication by Defra of the Animal Health and Welfare Strategy for the UK in 2004, leading to the Animal Welfare bill, and as a result of concerns for the horse industry in England following the introduction of the Act prohibiting hunting with hounds, there have been two strategy documents which offer hope for the future. These reports are: *Joint Research on the Horse Industry in UK*, produced by the British Horse Industry Confederation and Defra, and the *Health and Welfare Strategy for the Horse* produced by the British Veterinary Association and Defra. As a result, there is recognition that opportunities are being lost through the fragmented nature of the UK horse industry in terms of selective breeding, and this has been addressed for eventers by the production of genetic evaluations by the British Equestrian Federation through its breeding arm, British Breeding (Kearsley *et al.* 2006). Secondly, a passport system has been introduced, which will lead to the better identification of animals, so discouraging the indiscriminate breeding of horses and ponies of low quality or genetic merit. One requirement of the passport system is that the document should accompany the horse when it is used for breeding, thereby ensuring correct identification of the animal at that time. The legislation requires the establishment of a National Equine Database from information on passport applications, which will be a valuable tool for improvement in breeding, and, as a result, in the health and performance of competition horses and ponies in the UK. It has also been agreed that artificial insemination can be carried out safely by appropriately trained personnel, and an Exemption Order under the Veterinary Surgeons Act 1966 is currently under consideration to allow this procedure to be performed by people other than veterinary surgeons. These innovations should go some way towards redressing the historical lack of professionalism on the issue of breeding horses for sports in the UK, and help to provide structures which are taken for granted in some other European countries (e.g. Germany). They will, however, have little impact on hill breeds where animals run as herds and the assumption is that the alpha male is the sire of the foals. Although there is some indication that the herds do not mix, parentage cannot be assigned accurately on this basis. Molecular markers have not in the past been widely used in equine breeding, but work is currently underway to map the equine genome. Supported by the Horserace Betting Levy Board and the Childwick Trust an international collaboration (including in the UK the Animal Health Trust and the Royal Veterinary College) has been established aiming to produce a low resolution genetic map for the horse. The current genetic linkage map comprises approximately 800 markers and, excitingly, the horse genome is currently being sequenced. This development should accelerate research progress in horse breeding and permit the development of genetic screening tests for inherited diseases and their use to breed healthier horses and ponies. For example, the molecular defects underlying some of the important inherited diseases in horses (hyperkalaemic periodic paralysis, severe combined immunodeficiency syndrome, lethal white foal syndrome and junctional epidermolysis bullosa) have been identified. Severe combined immunodeficiency affects 8% of Arabian horses (Ding *et al.* 2002) and a diagnostic test which will identify carrier individuals is available. An improved cytogenetic test based on chromosome painting for the X0 sex chromosome abnormality, which renders affected mares sterile, has also recently been developed, which will improve the identification of these individuals. Note added in proof: there is now a complete sequence for the horse genome.

FUTURE NEEDS IN ANIMAL BREEDING AND GENETICS

Genetic improvement in livestock has a truly amazing history, with the *Journal of Animal Science* having been the world's premier publication venue for many of the major scientific innovations that have occurred since 1908 (Hohenboken, 2009). Although the majority of advancements in the past 100 yr have been made using quantitative genetics approaches, in recent years it has been nearly impossible to miss seemingly daily news reports about exciting discoveries in the relatively new field of molecular genetics and genomics. While most of these reports have focused on the unraveling of the human genome and its implications for human health, there has been significant spillover into plant and animal agriculture as well. At times over the past 15 yr, it has seemed that this new and exciting field would hold all of the immediate answers to breeding improved beef and dairy cattle, swine, sheep, and horses. Today we have draft genome sequence assemblies of the bovine and equine genomes completed and made publicly available along with a host of rapidly developing diagnostic tools, with the same now underway for the porcine and ovine species. Such transformational technological developments, coupled with several major societal shifts that are predicted to result in changes of epic proportions for the livestock industry and its producers, make the centennial anniversary of American Society of Animal Science (ASAS) an ideal time to consider needs in the field of animal breeding and genetics to fully address these challenges.

GENETIC THEORY TO SELECTION AND MATING SYSTEM TOOLS

The first half of the 20th century was an immensely prolific time in agricultural science. Arguably, the most dramatic discoveries were actually in the fields of genetics and statistics. During the 1920s and 1930s, the field of population genetics came of age, primarily as a means of quantifying and describing the writings of Darwin from the late 1800s. The emerging leaders of this field helped to describe the concepts of genes, gene loci, chromosomes, and cellular reproduction. They were also instrumental in establishing the field of biometrics—statistics as applied to biological phenomena. These early statisticians developed much of the underlying theory used broadly in science today. At the same time there were pioneering scientists who had the foresight to develop populations of livestock upon which they began to practice long-term selection and inbreeding.

Scientists also made what seemed to be an unrelated but extremely valuable discovery in plant genetics during this same period with the observation that when 2 unrelated lines of germplasm were crossed (hybridized), the resulting crossbred progeny had better performance than the expected average of the parents. The concept of heterosis between mildly inbred lines was born—and with it the seed industry and crop agriculture was revolutionized. At the time, livestock breeders did not see any great benefit from this phenomenon, but that would dramatically change later (Cundiff, 1970; Dickerson, 1973).

The era following World War II was a particularly exciting time for livestock genetic improvement, as it was in many fields. The 1940s saw some of the greatest minds to ever grace the study of livestock genetic improvement at their prime. J. L. Lush, who many refer to as the modern-day father of animal breeding (Dickerson, 1973; Dickerson and Willham, 1983), was busy defining, with coworkers Hazel and Dickerson, the concept of the "selection index" and "breeding value"(Hazel, 1943; Lush, 1947). The field of biometrics had matured to the point where it was now possible to determine from experimental populations that performance for traits affecting production could be measured, and that many of these traits appeared to be heritable.

In 1953, Watson and Crick presented for the first time in the scientific literature the molecular structure of the genetic code—DNA (Watson and Crick, 1953). Combined with the theories of genes and heritable variation of traits, it was now possible to visualize how these genetic differences at the gene level might one day be exploited for genetic improvement.

Also, in the 1950s, 2 significant events occurred that would permanently change the nature of livestock breeding. The first was that AI techniques matured to an adoptable level for cattle breeders, especially dairy producers. Coupled with the earlier institution of the Dairy Herd Improvement programs of USDA, volumes of data began to accumulate matching pedigrees to milk production records. At the same time, computing technology was beginning to surface as a usable tool, even though it was rudimentary to what we now have today. Dairy cattle breeders had enough foresight, however, to understand the power of coupling quantitative genetics theory to AI; as a result, genetic evaluation as an applied science was born. Furthermore, the plans of the dairy industry did not go unnoticed by other livestock breeding communities, and performance data recording quickly became mainstream practice in all of the species under various models, often at the level of breed registries and societies. Several organizations were also established in the industries to develop uniform guidelines for performance recording, including the Beef Improvement Federation, National Swine Improvement Federation, and the National Association of Animal Breeders. These organizations were provided critical leadership by animal breeders who were leaders in the field and in ASAS.

The late 1960s and early 1970s was the next period of great change in livestock breeding and genetics. Two things occurred somewhat simultaneously: the importation of germplasm from several lines and breeds and the next generation of computing technology coming of age. As a result of their faster growth rates, size, and altered body composition, several of these breeds quickly took a strong foothold in the seedstock industry. Breed associations and initial breeding companies took advantage of developing computer technology in the 1970s to deploy the first across-herd genetic evaluations using sire-maternal grandsire models, and the initial era of breeding value estimation across large-scale populations was born. These efforts built upon selection index theory and its applications employing mixed-model equations, first suggested by Henderson (1959, 1973, 1975) and further deployed by Willham (1979) among many others.

Additionally, producers discovered that hybrid vigour was indeed possible—and very economically beneficial—when many of the new breed lines were used in systematic crossbreeding programs (Dickerson, 1970, 1973). One of the most widely noted efforts in this area was developed and carried out at the USDA-ARS US Meat Animal Research Centre (MARC) through programs like the Germ Plasm Evaluation program led by Larry Cundiff and the Germ Plasm Utilization program led by Keith Gregory. Since 1968, this collective effort has produced the fundamental body of knowledge now used worldwide to understand genetic variation within and between breeds, and how to effectively use it in beef cattle production (Gregory and Cundiff, 1980). Similar programs and results have also been produced and published in the Journal of Animal Science for sheep and swine through the US MARC efforts.

The 1980s was a true time of transition for livestock breeding. Computing technology had matured to the level where statistical methodology could be applied to large pedigreed performance databases—so-called BLUP methodology. Scientists successfully worked out the reduced animal model, allowing the inverse of the numerator relationship to be solved (Quaas and Pollak, 1980) and were successful in using these methods to compute for the first time what we now know as EPD and predicted transmitting ability within breeds. These new genetic evaluation tools were significantly more powerful than the previously used breeding value estimates from the sire/ maternal grandsire model approach. In the time since, we have become the beneficiaries of continual refinement in genetic prediction methodology, including more accurate predictions as well as a plethora of new traits exhibiting useful levels of genetic variation (Koots et al., 1994a,b) added to the evaluation pipeline.

GENETIC VARIABILITY OF TRAITS

The objectives of this study were to estimate direct heritability of 30 conformation and performance traits evaluated during 100-day stationary performance test and to predict the genetic effects in Polish Warmblood stallions. Inbreeding coefficients were estimated as well. Moreover, phenotypic and genetic trends were derived. The data were extracted from the database of the Polish Horse Breeding Association. The analysis included 494 warmblood stallions, which performed during 100-day test in the years 2002-2008 at two Polish Training Centres. Pedigree data comprised 8512 individuals. Restricted maximum likelihood was employed to estimate parameters under an additive genetic model (including fixed effects: age of stallion, breed, year and place of performance test). Nonzero inbreeding coefficients were estimated for 88 stallions with performance records as well as for 458 unrecorded ancestors. Average inbreeding level for the stallions with records was 0.29%, whereas for all inbred individuals – 1.69%. The average completeness of the pedigrees, expressed as Cassell coefficient, for the recorded stallions was 42.47%. The heritability estimates varied from 0.14 (character) to 0.87 (total index). Relatively high heritabilities were found for jumping-ability traits. In general, the genetic trends for studied traits were negligibly positive.

Calving records from the Animal Breeding Centre of Iran collected from January 1990 to December 2007 and comprising 207106 first calving events of Holsteins from 2506 herds were analysed using linear sire models to estimate genetic trends for age at first calving (AFC) and milk traits. Genetic trends were obtained by regressing yearly mean estimates of breeding values on year of birth. In general, there were decreasing genetic trends for AFC, fat percentage and protein percentage over the years but there were increasing genetic trends for milk yield, mature-equivalent milk yield, fat yield, mature-equivalent fat yield, protein yield and mature-equivalent protein yield over the years.

On the other hand, there was a decreasing phenotypic trend for AFC but estimates of phenotypic trends were positive for milk yield and compositions over the years. It seems that the decline in calving age in this study over time resulted primarily from increased turnover rate and increased culling of heifers that failed to get pregnant. On the other hand, increasing trend for yield traits over time in this study indicated that Iranian dairy producers were successful in choosing progressively better semen and sires from imported and local sources over the years.

The main objective was to locate QTL and estimate the proportion of total genetic variance attributable to quantitative trait loci (QTL) for production index traits and the udder health index identified on six Bos taurus autosomes in the Danish Holstein dairy cattle population. Data were obtained from a granddaughter design of 20 sire families with a total of 1869 progeny tested sons. The number of sons per grandsire ranged from 20 to 284, with an average family size of 93.5. Indexes of the estimated breeding values were obtained for the milk production traits and for the udder health index from the Danish Agricultural Advisory Service database. A random-QTL model was applied to incorporate marker information into parameter estimation for each single QTL. The procedure allowed us to detect new QTL on BTA3, BTA16 and BTA28 and to estimate the proportion of total genetic variance attributed to different QTL on a total of six Bos taurus autosomes for the udder health index and yield index traits in the Danish Holstein population. Variance estimates vary between 2 to 58% of the total variance for different QTL and seem to explain a substantial part of the variance at certain positions of the cattle genome. The results are discussed against the background of the failure of marker-assisted selection (MAS) and the recent availability of large panels of single nucleotide polymorphisms (SNPs) that have improved the search for mutations underlying variation in complex traits resulting in modern genomic selection.

This study was carried out to investigate a new quarter-individual milking system called MultiLactor (Siliconform GmbH, Türkheim, Germany). The MultiLactor enables milking on quarter level basis with low vacuum (37 kPa), sequential pulsation and periodic air inlet. Within the same dairy farm, the influence of this quarter-individual milking system (MULTI) on milkability traits was compared with a conventional milking system (CON). CON was equipped with a conventional milking cluster and used alternating pulsation. Vacuum level was adjusted to 40 kPa. For the study, 84 Holstein Friesian cows were randomly selected and uniformly divided into two herds.

During the 30-week survey, the milk flow curves were recorded every other week by using a LactoCorder (WMB, Balgach, Switzerland). Significant differences ($P<0.05$) between both milking systems were found for all milk flow traits, except for milk yield and decline phase. Concerning the incline (tAN) and plateau (tPL) phase, large differences existed between MULTI and CON. The estimated value of tAN calculated for MULTI (29.4 s) took only half of the time when calculated for CON (56.4 s). The estimated value of tPL at CON was reduced by 1.43 min (35%) compared to MULTI. Milking process at MULTI (8.49 min) took longer time than for CON (7.43 min). From the study, it was concluded that the effect of shorter tAN in the quarter-individual milked cows may be related to additional prestimulation by an actuator. In contrast, the longer milking time in MULTI is possibly caused by lower vacuum level and periodic air inlet.

The objective of this study was to examine semen quality parameters for the Pøeštice black-pied breed over a test period of 8 years while considering the potential effects of collection month and boar age. Ninety-nine ejaculates were collected using the gloved-hand technique from healthy and fertile mature boars from selected farms. Ejaculate volumes were relatively low because the boars were accustomed to natural mating. Sperm motility, sperm concentration, percentage of morphologically abnormal spermatozoa (MAS), and sperm motility after 24 h of storage in Androhep extender (dilution rate 1:1.5) were assessed. Significant differences were found in sperm concentration and MAS rate in relation to collection months and boar age in the monitored years ($P<0.05$). A tendency for MAS to increase with monitored years was observed. Significant differences in sperm motility and motility after 24 h of storage were only observed in relation to collection months ($P<0.05$). Results of this study detected effects due to collection month and boar age on boar semen quality during the monitored years.

The objective of the present study was to determine the polymorphism in the prolactine receptor (PRLR) gene in Chios, White Karaman and Awassi, which are native sheep breeds in Turkey. By means of PRLR gene sequence homology between sheep and humans, two primer pairs were designed for polymerase chain reaction (PCR) amplification within intron 1 and exon 10 of the PRLR gene in sheep. A total of 160 amplicons (99 for intron 1 and 61 for exon 10) were subjected to DNA sequence analysis. For intron 1, 6 different haplotypes were determined. For exon 10, 7 different haplotypes were obtained. Some variations determined for exon 10 (g.14A>T p.Q14L; g.160G>A p.D160N; g.166G>A p.E166K; g.167A>T p.E167V; g.176A>T p.H176L; g.206G>A p.S206N; g.208G>A p.G208R) led to changes in the amino acids, but no amino acid changes were determined in g.2A>T, g.81A>G, g.138A>G, g.186C>T, g.207T>C. It was noted in particular that White Karaman and Awassi were similar to each other in both PRLR exon 10 and intron 1 haplotypes, whereas the Chios breed had a different variation.

The aim of the presented study was an estimation of the zinc bioavailability derived from amino acid complexes with methionine, lysine and glycine in growing lambs. 48 lambs, Polish Merino × Romanowski × Charolaise crossbreed, at age about 10 weeks and average body weight of 20 kg were randomly divided into 4 experimental groups. Animals were kept collectively and fed with concentrate and hay in amount: 0.7-1.2 and 0.3-0.5 kg/day/head, respectively. After rearing period 6 rams from each group with average body weight of about 30 kg were divided into digestibility-balance experiments. At the end of the experiment the blood samples were taken, then from each group 8 lambs were chosen, killed and during dissection tissue samples were taken (liver, pancreas, kidney and rib bone).

In tissue samples as well as in fodders and excrements the content of zinc was determined. Obtained data in digestibility experiments allowed on calculation of apparent absorption and retention of zinc and the level of Zn determined in tissues' samples were used to the estimation of bioavailability of zinc from different ones applied in experiment sources. Apparent absorption and retention of zinc were higher (Pd"0.01) in lambs receiving in mixtures amino acid complexes of zinc than in animals from the control group. The higher content of zinc in soft and hard tissues of lambs which received organic forms of zinc might indicate better assimilation and bioavailability of zinc from these forms than from oxide. Among tested organic forms of zinc, the zinc-lysine complex was characterized as having the most advantageous property.

The aim of this study was to assess the seasonal trend of the serum protein content of sheep and goats. The tested animals were six female goats (Maltese breed) and six female sheep (Valle del Belice breed). All animals were clinically healthy and not pregnant or lactating before or during the study. On blood samples, collected through an external jugular venipuncture every

30 days for 12 months, electrophoresis was performed using a semiautomated AGE system and then electrophoretic curves with the relative protein concentrations were analysed. One way for repeated measure analysis of variance (ANOVA) was applied to determine the effect of time and by means of cosinor rythmometry, mesor (mean level), amplitude (half the range of oscillation) and acrophase (Φ, time of peak) were determined. The results showed a seasonal rhythm on Albumin and Alumin/Globulin ratio for sheep and goats, with different acrophases, winter for goats and spring for sheep. A seasonal rhythm was shown also in Alpha 2 globulins by sheep and in Beta globulins by goats. The difference in the acrophase can be attributed to a different production pattern of melatonin in goat, so the liver production of albumin is major during the winter that has a longer scotophase.

A total of 238 Pannon White growing rabbits were slaughtered at 74, 84 and 94 days of age, at an average body weight of 2.53, 2.84 and 3.15 kg, respectively. Within each age group five body weight categories were formed in such a way that the weight difference between two neighbouring groups was equally 0.3 kg. The design of the experiment permitted the separate examination of the effects exerted by age and body weight on carcass traits and on meat chemical composition of the left hind leg (HL) and that of m. Longissimus dorsi (MLD). Dressing out percentage significantly increased with the increase of age in all body weight categories, as well as with the increase of body weight. The effect of age on the ratio of the fore part of the carcass depended on the body weight: significant difference was found in lower body weight categories. The ratio of the intermediate part of the carcass was not affected by age but it increased in rabbits of higher body weight. The ratio of the hind part of the carcass increased in older animals, while it decreased in rabbits of higher body weight, and this decrease was higher in older rabbits. The ratio of the perirenal fat to reference carcass significantly decreased with age in all body weight categories, but it significantly increased in rabbits of higher body weight. With the advancement of age the moisture content of the HL increased, while that of the MLD did not change. With the increase of body weight, moisture content decreased in both the HL and in the MLD. The fat content of both HL and MLD decreased in older animals, while it increased in rabbits of higher body weight.

The aim of the study was to provide information on the genetic variability of the Hungarian Bronze turkey gene reserve population and its difference from the Broad-breasted turkey, and offer guidance and proposals for its future conservation strategies. Altogether, 239 Hungarian Bronze turkeys from 10 strains and 13 Broad-breasted turkeys as a control population were genotyped for 15 microsatellites. All loci were polymorphic with the average number of alleles per locus 3.20±1.146 in the Hungarian Bronze turkey. The mean expected (Hexp) and observed heterozygosity (Hobs) were not different (0.392 and 0.376, respectively) in the overall population, and similar values were obtained for hens and bucks and among hen strains. Inbreeding coefficient (FIS) and Shannon index (I) indicated that there was low inbreeding within hens and bucks. Our results confirm that the genetic diversity in the Hungarian Bronze turkey population has been preserved by the rotational mating system. Differences between the Hungarian Bronze turkey and the Broad-breasted turkey populations were determined. Nei's unbiased values clearly indicated that the two populations are highly genetically differentiated.

The membrane attack complex (MAC) of the complement system is known as a natural immune mechanism of hosts against invading pathogens. This study was undertaken to structurally characterize and computationally analysed the 5′-region, i.e. the downstream promoter region and the 5′-UTR nucleotide sequence of the porcine C8A, one of the components of the MAC. Sequencing approximately 950 bp of the 5′-region and analysing in silico revealed the transcription start site, the TATA-box, the CAAAT-box, and the GATAAbox. High identity was found in range of 37-74 % among the sequences of pig, human, cattle, and mouse. Transcription factor binding sites, such as NFκB, Oct-1, HNF1, CDP, and C/EBP, showed high conservation between vertebrate animal species, especially between human and mouse, or pig and cattle. Seven single nucleotide polymorphisms were detected in the breeds Hampshire, Duroc, German Landrace, Large White, Pietrain, Berlin Miniature Pig, and Muong Khuong. Nucleotide exchanges could cause the generation of new binding motifs, which may affect the expression of the porcine C8A, particularly the C/EBP regulation of the porcine C8A gene-as described in the human C6 and C7 promoter.

PRECISION ANIMAL BREEDING

We breed animals for four principal reasons: as sources of usable products or services; for medical or scientific research; for aesthetic, cultural or ethical considerations; and as pets. The first leads to animal husbandry and livestock breeds of domesticated species kept for food, fibre and other services such as transport and power; the second provides laboratory animals of defined genetic lines, including animals with gene knockouts; the third encompasses breeding for conservation; and the fourth leads to companion animals used for pleasure or recreation. Objectives of sustainability should apply in each case. Precision animal breeding should be used whenever animals are bred *for a purpose*; that is when they are bred for a particular use, environment or market. It should therefore apply when animals are bred for any of the four reasons described above. This is appropriate as the techniques used are largely generic. Our aims in this paper are: (i) to review the methods currently available, (ii) to indicate how they have developed over time to become more precise, (iii) importantly, to show how generic technologies are giving opportunities to substantially improve precision, and (iv) to show what needs to be done to deliver these opportunities. In many cases, the breeding of animals is controlled or influenced by legislation, or by national and international bodies. It is not possible to review precision animal breeding without reference to either of these factors, or the sizes of animal populations or their economic and cultural significance, and for this reason it has been necessary to set the science within a wider background. On the other hand, we have had to limit the range of species considered and will therefore concentrate on terrestrial vertebrates.

The twentieth century marked a turning point in our relationship with other species from which there is no way back. The human population explosion led to widespread competition with other species for agricultural land, and many species became extinct or are now threatened with extinction primarily as a consequence of these activities. We also began to recognize how our industrial culture influences our environment not only locally but also at a distance through atmospheric pollution (e.g. acid rain), and globally through climate change.

We accept that where animals are already within our care, we are responsible for their quality of life. We accept that the activities of man affect all the living things with which we share this planet. But we are slow to realize that as a result we have a duty to care for all living things. This duty extends to the breeding of animals for which we are responsible.

It is in this context that we propose that precision animal breeding should set the following goals.

1. To increase the scope and precision of predictions of the outcomes of breeding decisions (G1).
2. To avoid the introduction and advance of characteristics deleterious to animal well-being or, more generally, the well-being of the species (G2).
3. To manage genetic resources and diversity between and within populations in accordance with the principles set out in the Convention on Biological Diversity.

While these goals will be discussed in more detail later, we shall advance some initial arguments for their form. The first of these goals is clear in the context of precision breeding! However, we have the potential through genomic technology to be more precise in directing the outcomes to be closer to what we desire, both directly in what we select for and indirectly in reducing the chance of unforeseen consequences. The second goal recognizes not only our responsibility for quality of life in populations that are managed, but also some of the ethical conflicts that arise in conservation activities, where an aim for a captive population is to support or re-establish a wild population, and where domestication through adaptation to captivity may be harmful. The third goal recognizes the force of the arguments clearly articulated in the Convention on Biological Diversity, which apply to all our genetic resources both managed and unmanaged.

To meet these goals for precision animal breeding there is a requirement to establish:

1. breeding objectives in the context of the environment in which the progeny are to be kept (R1);
2. a robust parameterized model of the extent of genetic variation, for example heritabilities of traits selected and genetic correlations among significant traits, together with estimates of breeding values (BVs) for significant traits for individual animals (R2);
3. an understanding of gene expression from genotype to phenotype, including the molecular basis of traits (R3).

In practice, for the vast majority of species all this information is seldom available. At best we meet R1 and R2 and a small component of R3 in a limited number of commercially valuable populations. It should be an urgent goal to extend that number as rapidly as possible in order to meet the goals of precision animal breeding, as once desirable population characteristics are lost or conversely deleterious characteristics are introduced, the desired position can be slow and costly to retrieve. Furthermore, in view of the international trade in genetic resources, this need for information must be considered at an international level. There is a continuum between the genetic structures of small, endangered populations and those of widely used livestock species, and in many cases the same issues affect both. For instance, both the Holstein dairy cow and the Arabian oryx (*Oryx leucoryx*) are affected by concerns over rates of inbreeding. In the case of the dairy cow this results from the widespread use of a limited number of desirable bulls through artificial insemination. However, the degree of inbreeding in these two species differs markedly: for the dairy cow, *F*H⁻2.5% in the UK to a 1950 base (Kearney *et al.* 2004) whereas for the oryx in 1985 *F*H–60% to a 1970 base (Mace 1988). Nevertheless, the methods used to trace an animal's evolutionary history may be applied to both rare and widely bred species: for instance microsatellite markers have been used to understand the evolutionary history and current genetic status of Arabian oryx (Marshall *et al.* 1999), South American camelids (Bustamente *et al.* 2003) and the Jersey cow (Chikhi *et al.* 2004). The same procedures are therefore available for application to all managed species, in both small and large populations. Whether it is appropriate to apply them depends on the question being asked and the cost-effectiveness of the procedure, given that there may be other calls on resources.

THE STATE OF THE ART

The state of the art of animal breeding is based on integrating aspects of many sciences and technologies, with the degree of sophistication varying considerably between the different areas of interest and activities. The key scientific inputs range from genetics (both quantitative and molecular), statistics and computing science, information technology, and the physiology and endocrinology underlying reproduction and fertility. The integration concerns the objectives of the breeding, testing, recording, evaluation, selection and mating of breeding animals.

The objectives of the different areas of breeding activity vary widely. In livestock, well-defined objectives (R1) are necessary for success and will usually evolve over time in relation to societal concerns. Objectives have been for productivity with higher output per unit input but, while these objectives are still important, there has been a notable shift towards tackling issues related to reproduction, well-being and longevity. The ability to tackle the latter objectives is currently constrained in many parts of the world, including the UK, by a combination of poor information on these traits owing to the lack of routine recording coupled to the poor genetic 'signal' within this information. In conservation activities the objective is primarily to increase population size and conserve genetic variation; however, where conservation management includes breeding for release, an important genotype-by-environment interaction arises since breeding in captivity must address the objectives of non-domestication and fitness for wild environments (Gilligan & Frankham 2003).

To illustrate the state of the art we shall describe dairy breeding as a reference and indicate how other sectors differ in technology and operation. The challenge with dairy breeding is that key traits, such as milk production, are sex-and age-limited, and this determines much of the structure of the industry. These sex-and age-limited traits can only be measured in the mature female, reducing the scope for selecting animals on their own performance and the speed at which selection can occur. This is particularly restrictive as it is the male that has the much higher reproductive rate, particularly since advances in reproductive sciences led to the introduction of artificial insemination some 50 years ago. Therefore, improvement systems have developed that are based upon progeny testing in which the genetic merit of candidate bulls is judged by a large number of daughters (100 or more in some cases) being milked on many private farms across the country. Conducting such a test efficiently is a challenge requiring efficient statistical design, the routine gathering of records on performance, and their collation and storage in databases for evaluation.

The estimation of genetic values for the bulls being tested, and for all other animals in the dairy herd, is made using these data and all the other records collected on dairy cows, using best linear unbiased prediction. This statistical technique is based upon detailed genetic models (R2) and accounts for the diverse environments of the thousands of different farms from which the data has been collected. The computational challenge is considerable: predicting BVs for millions of animals across thousands of farms, for hundreds of traits, requires major computational and statistical efforts. Agricultural breeding companies in the pig and poultry sectors are similarly developed operationally, although less reliant on progeny testing, and with the added benefits of the ability to manage the necessary size of population and the scale of recording from within their own resources. Additionally, several breeding companies use sophisticated selection techniques and mating procedures to minimize the loss of genetic variation in the process of generating improvement (Woolliams et al. 2002), a clear step to achieving G3 within commercial livestock. For some sectors, a limited number of genetic markers may contribute information to the selection process; where this is done the most valued markers are those that address traits such as disease or reproduction where other genetic information usually has poor accuracy. Other sectors such as sheep breeding are less well developed in this respect.

For companion animals and equids, the state of the art for breeding is centred upon the use of pedigree records and shows to exhibit examples of the breed and identify desirable parents, together with the occasional use of reproductive technology to increase the reproductive rates of such parents. In equids, there is a move internationally towards more reliable statistical evaluations based on performance and formal testing of young horses. Animals bred for science differ again. The populations of mice that are used are predominantly well-documented inbred strains and crosses derived from them. The use of transgenic technology and mutagenesis to produce novel phenotypes has increased significantly. Limiting the range of species considered here precludes mention of Drosophila, which continues to make a huge contribution to functional genomics, and zebrafish.

POPULATION GENETICS

The physical and behavioural characteristics of organisms are determined by the genes they inherit and the environment in which they develop. In terms of physical characteristics, this is widely accepted. Those doubting the importance of genetics in controlling behaviour should consider that behaviours are determined by physical characteristics of neuronal pathways and interactions (e.g. Kendler 2003).

There are approximately 30 000 genes coding for proteins in the nuclear genome of the human being (Southan 2004), which may be taken as a working estimate for other vertebrates. Genes are composed of sequences of nucleotides, and there are differences between individuals in the sequences of nucleotides within a gene known as *polymorphisms*. There are also differences between individuals in the degree to which genes are expressed due to polymorphisms in the regions controlling transcription or in other genes, which control transcription in *trans*. These differences between individuals account for the genetic differences between individuals, which together comprise the *genetic variation* in a population. Where genes determine quantitatively measurable characteristics, these characteristics are known as *quantitative traits*.

During fertilization, an individual receives one copy of each gene (with a caveat over sex-linked genes) in the gamete received from each parent, with the genes passed in discrete blocks called chromosomes. In humans, the number of chromosomes passed from parent to offspring is 23, and in cattle 30. The chromosomes passed by the parent will be a random choice among the pair of homologous chromosomes carried by the parent made during the process of gamete formation, called meiosis. However, the process of meiosis involves *recombination* in which segments of the two homologous chromosomes carried by the parent are exchanged. As a result, new gene sequences may arise. However, there are few recombination events relative to the total number of genes. In addition, variation can arise from new mutations within an individual that change a nucleotide or involve more complex deletions and rearrangements, and change the functional properties of a gene. If these occur within cells of the germ line, these mutations can enter the gene pool of the population and add to the genetic variation observed.

What is observed is the phenotype, which is determined by the sum of the genetic effects and the environmental effects. The latter is the accumulated impact of events in the life history of an individual and, possibly, its parents or even more distant ancestors (e.g. Benton *et al.* 2005). The primary clue for recognizing the existence of genetic variation is the finding of a degree of resemblance between relatives, especially when their development has occurred in different environments.

While we can observe genetic variation within a population, not all of it can be used to produce a change in the population mean; the portion that can be used in this way is called the *additive genetic variation*. It is knowledge of the extent of variation and its different portions that comprise a requirement

R2 for precision animal breeding. Knowledge of the precise mutations that underlie genetic variation is unnecessary in order to achieve improvement. For example, despite the doubling of milk yield since 1940 and the many genes that affect milk yield, no single mutation having a direct and verifiable effect on yield was known until 2000 (Andersson & Georges 2004). The better our understanding of the variation, advancing to knowledge of the important mutations underlying variation, i.e. including addressing requirement R3, the more capable we will be of achieving the goals G1 and G2 of precision animal breeding.

The superiority of the offspring from an individual parent when compared with the population mean leads to the concept of the *BV*, which is a measure of the additive component of the parent's genes and its suitability for *selection*. Most quantitative traits are determined by many genes acting in concert (i.e. are *polygenic*), and so an individual's BV cannot be considered the property of a single gene, but of many genes in combination. Breeding schemes are a long way from achieving full precision so man-made selection (*artificial* rather than natural selection) is made on an *estimated breeding value*.

Genetic Correlations

Since some genes affect more than one trait, and since some genes, being located close to one another on the same chromosome, tend to be inherited together, there are *genetic correlations* between certain traits. That is, genetic improvement through selection for one trait may lead to a change (improvement or otherwise) in another.

This has been recognized since selective breeding began; Darwin, in *The origin of species by means of natural selection* states: Hairless dogs have imperfect teeth; long-haired and course-haired animals are apt to have, as is asserted, long or many horns; pigeons with feathered feet have skin between their outer toes; pigeons with short beaks have small feet, and those with long beaks large feet. Hence if man goes on selecting, and thus augmenting, any peculiarity, he will almost certainly modify unintentionally other parts of the structure, owing to the mysterious laws of correlation.

Genetic correlations have resulted in undesirable and unexpected side effects of selection in many species. In dairy cattle for instance, selection for yield has resulted in progressive loss of fertility, so that while yield has increased annually in the UK by approximately 90 kg per lactation for the past 20 years, rates of fertility (measured as per cent of animals conceiving to first service) have declined during this period by about 1% per annum (Royal *et al.* 2000): the genetic correlation between yield and calving interval in UK Holsteins is 0.27, and between yield and days to first service, 0.67 (Wall *et al.* 2003). To counter this effect, *fertility indexes* have been introduced initially in Nordic countries and more recently in the UK (Flint *et al.* 2004) to promote those bulls with high yielding, fertile daughters which, though fewer in number, are present in the population. Examples of other genetic correlations that have adversely affected breeding outcomes include the correlation of growth with egg production in poultry (Pirchner 1986) and fleece weight with fertility in sheep (Fogarty 1995). It is only by understanding the mechanisms that lead to genetic correlation, through advances in knowledge of gene expression, that we can be truly proactive in achieving goals G1 and G2, rather than only responding to what is observed.

Molecular Genetics, Genome Sequencing

The revolution in molecular genetics, and particularly genome sequencing, has already provided benefits for animal breeding, but in comparison with what the future holds, our present tools will undoubtedly be seen as primitive. At the time of writing, complete genome sequences are available for a number of species, genome sequences for the chicken, cow, horse, mouse and chimpanzee are either completed or nearing completion, and single nucleotide polymorphism (SNP) libraries for these species are growing rapidly. This information will underpin most of the developments in livestock breeding and breed management during the coming two decades.

Molecular Markers, Microsatellites and Single Nucleotide Polymorphisms

The value of molecular information in assisting breeding decisions has already been demonstrated, in particular through the use of marker-assisted selection as well as for monitoring population structure, and for information on the history and development of populations. Early examples of gene discoveries in livestock that were subsequently used for marker-assisted selection were: a mutation causing malignant hyperthermia in pigs (Fujii *et al.* 1991); a microsatellite closely linked to the mutation causing weaver disease in cattle (Georges *et al.* 1993); a marker for increased litter size in pigs (Rothschild *et al.* 1996); and a deletion leading to double muscling in cattle (Grobet *et al.* 1997). The use of DNA markers for monitoring intra-breed structure has been verifiably illustrated by Chikhi *et al.* (2004) in their study of Jersey cattle on the Isle of Jersey.

The primary route of development for DNA marker technology has been from restriction fragment length polymorphisms (RFLPs; which arise from loss or gain of sites where genomic DNA is cut by restriction enzymes), through microsatellites (arising from variation in the lengths of repeated sequences of satellite DNA) to SNPs (single point variations in nucleotide sequence).

Among other marker types, SNPs share the benefits of being: (i) widespread throughout the genome and so close to coding regions, (ii) co-dominant hence heterozygotes can be identified, and (iii) amenable to the use of PCR to amplify the signal. The use of microsatellites has been widespread over the last decade, with advantages over RFLPs in the multiplicity of alleles at a single locus and a greater repeatability, although reproducibility across laboratories is still a problem.

The technology surrounding SNPs has recently improved by orders of magnitude, so that assays can be scaled and automated, resulting in a low cost that outweighs the disadvantage of being predominantly biallelic. Blott *et al.* (2003) estimated that it is necessary to have five or six closely linked SNPs to replace the information contributed by a typical microsatellite, but given the current approximately 100-fold difference in cost per genotype, the benefit is clear. One outcome of such a reduction in cost is that a cattle genome can be densely marked with 50 000 SNPs at a current (2007) cost of approximately £200; this is a small cost when compared with the cost of breeding and testing a dairy bull for example.

Changes in nucleotide sequences within coding regions are termed 'silent' if they make no change to the amino acid coded, but if silent they may change transcription rates or transcript turnover, thereby affecting protein levels. However, the complexity of control of DNA transcription is still poorly understood. An example of this complexity arising from a single SNP is the polar overdominance in the ovine callipyge gene (Freking *et al.* 2002).

Mitochondrial DNA

Mammalian mitochondria contain a small circular DNA plasmid of 16.5 kb which codes for 37 genes required to be expressed within the inner mitochondrial membrane. The mitochondrial genome evolves 17 times faster than nuclear DNA, probably due to lack of DNA repair mechanisms. As a result, the mitochondrial DNA sequence can be used to monitor evolution on a shorter time scale than is possible with chromosomal DNA.

Mitochondrial genome polymorphisms are therefore frequently used to analyse population structure and demographic history. Mitochondrial DNA is haploid, and so each individual has a single haplotype.

It is maternally inherited as a result of the limited contribution to the zygote of mitochondria from sperm. This limits its use in relation to domestic species where gene flow through the male line forms an important determinant of evolution and population structure, for instance through artificial insemination, but provides the advantage that it allows introgression through the female line to be distinguished from that through the male. The value of mitochondrial DNA for animal breeding is that it allows an understanding of population history and structure in time and space. For instance, a low level of mitochondrial DNA polymorphism within a species suggests it has survived a reduction in population size or bottleneck, whereas a high level of variation is characteristic of a large and well-established population. Mitochondrial DNA is extremely valuable in resolving important taxonomic questions when distinguishing subspecies and in identifying evolutionary significant units. Mitochondrial DNA bar coding has been suggested as an aid to assessment of biodiversity (Hebert *et al.* 2003), but this has been questioned, particularly in species subject to parasite infestation or with a high incidence of symbiont infection (e.g. arthropods; Hurst & Jiggins 2005). The general conclusion from large-scale analyses of livestock populations, which are perhaps the most informative for this purpose, suggests that variation in maternal lineages explain at most a small fraction of the variation in traits of commercial interest (e.g. Roughsedge *et al.* 2000*a*,*b*).

Epigenetic Effects

Certain heritable characteristics are encoded in DNA by covalent modifications to chromosome structure rather than the nucleotide base sequence of genes. Examples of these modifications are the acetylation of histones in nucleosomes, which alters the ability of the transcriptional enzymes to open and copy the DNA, and the methylation of cytosine residues in DNA, altering their protein-binding characteristics. These alterations in chromosome structure are responsible for genome imprinting, which determines whether a copy of a gene received from the mother or father is transcribed. This process is responsible for genetic diseases such as Huntingdon's chorea through revealing a deleterious mutation that remains uncompensated. Imprinted genes are also involved in the expression of the callipyge mutation (Freking *et al.* 2002) in sheep.

Genetic Evaluation

By knowing BVs, changes can be made in traits without knowing what genes are responsible, and this is what has been exploited throughout domestication: the heritability of characteristics and the similarity of relatives. The development of this process has produced the advanced methods available today, of which BLUP is the epitome for the present.

The introduction of accurate methods for the estimation of BVs has been one of the major successes in precision animal breeding. The need for improved statistical techniques was recognized following the development of artificial insemination during the 1940s, and the introduction of progeny testing for dairy bulls in the 1950s. The method used at that time was *contemporary comparison*, which used production data on a bull's daughters during their first lactation, compared with other cows in the same herd during the same year and season. Effects were then combined across herds.

Contemporary comparison recognized the importance of environmental effects, but failed to take into account differences between herds in genetic merit. It also assumed that all animals other than the daughters of the sire under test were unrelated, and this became increasingly invalid following introduction of artificial insemination. To overcome these difficulties, a method was required which allowed for a wider variety of environmental variables, and took account of the genetic relationships between the animals sampled. These requirements were met by the introduction of BLUP developed by C. R. Henderson in 1949 (Henderson 1975).

This statistical method uses matrix algebra to solve the large numbers of simultaneous equations generated for estimating BVs and for identifying environmental effects simultaneously. Implementation of BLUP required considerable computing power to solve the matrices generated. In particular, the complexity of the genetic information used dictates the amount of computer space or time required. In dairy cattle breeding, sire model BLUP evaluations were first used in the US in the early 1970s, sire–maternal grandsire BLUP evaluations were introduced in the UK in 1979 and animal model BLUPs in 1992. BLUP is the current method of choice for evaluation in all sectors and is widely implemented.

Reproductive Technologies

The development of population genetics methodologies and their application to livestock breeding coincided with the development of reproductive physiology and the introduction of reproductive technologies during the last half of the twentieth century. These technologies include artificial insemination and embryo transfer, both of which have been used extensively in the international dissemination of genetic resources. The development of assisted reproduction techniques in human medicine has gone hand-in-hand with their application in animal breeding, advances in human medicine being dependent on studies in animal models.

Artificial Insemination and Multiple Ovulation and Embryo Transfer

Examples of the impact of reproductive technologies on the dissemination of improved genetics are to be found in the turkey and dairy breeding industries. Commercial turkey breeding is now dependent on artificial insemination, because the large size of the males of broad-breasted turkeys, which have been bred for body conformation, precludes natural mating (a failure to apply precision animal breeding). However, artificial insemination provides great benefits. Without semen cryopreservation and artificial insemination the dairy industry would not have developed in the way that it has, since the 1950s, through the international trade in semen and embryos, and these techniques have also had an impact, though less dramatically, on the beef and sheep industries. Since these methods were developed, we now have semen and embryo sexing and multiple ovulation and embryo transfer (MOET) technologies, somatic nuclear transfer and assisted reproduction techniques developed in human medicine, such as intra-cytoplasmic sperm injection. All these techniques, but particularly semen cryopreservation and artificial insemination, play a major role in the design of breeding programmes and in the dissemination of advanced genetics, although the use of nuclear transfer is limited at present. These technologies allow better identification of merit and increased selection intensity through increasing reproductive rate and better use of resources for testing.

The benefits of artificial insemination do not require its widespread use. For instance, artificial insemination makes *sire referencing schemes* possible, where a number of farms cooperate in a breeding programme, using the same reference sires. This has the advantage that by using a small number of semen donors, accurate genetic comparisons across farms can be obtained. These schemes have been extensively used in breeding sheep and beef cattle, and have also been used in the UK for red deer.

Transgenesis

Use of gene insertion and gene knockout techniques has been largely limited to the production of animals for medical research. An exception has been in the production of animals secreting proteins of pharmaceutical value in their milk (Colman 1996). The artificial modification of genes encompasses transfer of genes from one species to another, or altering the genome to remove or suppress the expression of existing genes. Within livestock breeding there are no transgenes within the gene pools of commercial populations at present, outside the experimental stations of various countries. In some countries, for example Norway, the use of transgenesis in livestock is prohibited on principle.

What is clear is that the difficulties of ensuring the necessary integration of the transgene into the genome (both in controlling expression of the transgene and integrating expression with the remainder of the genome) has been a challenge of questionable cost–benefit. Early efforts in this area (e.g. Wagner *et al.* 1983) clearly failed in relation to principle G2 that we advocate for precision animal breeding. However, future advances in understanding may lead to applications that satisfy all the principles listed and which will have a benefit sufficiently large to carry public opinion, for example in the prevention of endemic diseases and zoonoses.

Nuclear Transfer and Cloning

Besides its potential to contribute to the study of developmental biology, somatic nuclear transfer has two practical advantages: as a route to transgenic animals and stems cells, and as a means of conserving nuclear genomic material. Cloning by nuclear transfer has now been applied in many species: sheep (Dolly, the first from an adult cell; Wilmut *et al.* 1997); goats; cows; pigs; horses; a rhesus monkey; rats; mice; dogs; cats; rabbits; and gaur.

As a means of producing transgenic animals, the advantage of nuclear transfer lies in the modification and selection of nuclear donor cells in culture, before transfer of their nuclei into enucleated oocytes. Following transfer, factors in the oocyte cytoplasm reprogramme the donor nucleus as a result of which it reverts to totipotency. In this respect, the technique provides the advantages of embryonic stem (ES) cells in species other than mice, which is important because stem cells have not been derived for domestic species.

All the gene transfer methods applied to mouse stem cells can be used in cells for nuclear transfer, including random gene insertion, targeted gene replacement by homologous recombination and targeted gene insertion. Selection after transfection can be by toxin resistance, and changes can be monitored by standard molecular biology techniques before cells are used.

Despite some improvements in the success rate of nuclear transfer (Campbell *et al.* 2005; Lee & Campbell 2006), much still needs to be done to understand the limitations inherent in the process, and to develop more efficient procedures.

The advantage of nuclear transfer as a means to conservation lies both in the ability to freeze somatic cells and in its use to rapidly increase population sizes. Somatic cells are easier to obtain than gametes, although freezing/thawing regimes are better developed for sperm and eggs in a number of species than for other cells. Once transferred to oocytes, the population of cells represented by the conserved culture is rapidly promulgated. For instance, in cats, large litters can be obtained from the normal mating of individuals, both of whom were produced by nuclear transfer, and this methodology is applicable to rare species such as the African wild cat (Gómez *et al.* 2004).

Applications of nuclear transfer to agriculture include not only the preservation of valuable individuals and rare breeds, but also the rapid propagation of high-genetic-merit animals and animals of specific genotypes (Woolliams & Wilmut 1989). In this respect, nuclear transfer will benefit precision animal breeding by increasing the predictability of livestock performance (G1) and, perhaps paradoxically, fulfilling G3, although Woolliams and Wilmut highlight the risks associated with reducing local diversity through the use of nuclear transfer.

Value Added by Selective Breeding

A conservative estimate of the annual value of livestock production in Europe is €123 billion, and the annual genetic gain at the level of the producer is equivalent to 1.5% of that figure, i.e. €1.8 billion (FABRE-TP 2005). The annual research and development costs of breeding organizations, including collecting data for estimating BVs and carrying out breeding programmes but not product marketing, is approximately €150 million, a benefit to cost ratio of 10. Genetic gains are permanent and cumulative so that the gain made in one year will give benefits over all subsequent years without further intervention, and this increases the benefit to cost ratio cohort by cohort. This contrasts with vaccination strategies, for example, where the benefit from a vaccine requires repeated application, cohort by cohort. The livestock sector in Europe employs 3.5 million people and remains the largest sector in agriculture in terms of both employment and output value. Therefore, genetic progress is central to the success of this major industry.

Structure of the Industry

Progress is principally in the hands of a small number of large institutions. The costs associated with collecting information on economically significant traits, developing and maintaining extensive pedigree and other databases, using up-to-date statistical techniques and large computer systems are high, and as a result much of the expertise is concentrated in a few large commercial companies. For example, over 90% of global poultry breeding stock (layers, broilers and turkeys) is in the hands of two or three organizations selling to worldwide markets in each case. Similar situations apply for pigs and dairy cattle. On the other hand, there are a large number of small breeding organizations dealing with individual breeds, which are unable to benefit easily from the advances open to larger companies. This barrier can be overcome through cooperatives, and effective examples can be found in many European countries (e.g. the Genesis Faraday Partnership scheme in the UK).

Aims for Genetics Within Livestock Industries

The general aim of the livestock sector is to meet the aspirations of the world's population for increased availability of animal products in a sustainable manner while ensuring food safety, animal welfare and the maintenance of rare and specialist breeds. The expected worldwide increase in consumption of animal products for the next decade is 7% annually. The availability of marginal land suitable for producing this increase is limited, and hence there is a need to produce more from the same resources (Food and Agriculture Organization 2000). For genetics to contribute half the required acceleration would mean a doubling in genetic gain. This illustrates both the need and the opportunity for livestock breeding, and placed alongside the requirements for sustainability, the need for precision animal breeding. Sustainability in livestock production implies meeting production targets while ensuring targets are also met for environmentally significant outputs, human feed efficiency, animal health and welfare and maintenance of biodiversity, in both farmed livestock and wild species affected by animal husbandry.

Modern agricultural practices often reduce, rather than increase, genetic diversity in domestic animal populations. Selection for desirable traits and the rapid dissemination of genetic material through populations by artificial insemination and embryo transfer, including international trading in genetic material, all tend to reduce genetic diversity. Furthermore, genetic diversity is lost through the reduction of population sizes in rare breeds, or in some cases through the loss of the breeds themselves. The genetic diversity being lost in this way is not fully characterised. Such trends are contrary to goal G3 and are explored further below. Within breeding schemes goal G3 can be addressed using selection procedures that explicitly manage genetic variation whilst maximising gain through controlling the rate of inbreeding (e.g. Meuwissen 1997).

'Pharming' and Xenotransplantation

Most pharmaceuticals currently available are small molecules. Their chemistry is accessible and they are relatively cheap to produce. In contrast, many therapeutic compounds are proteins, and advances in functional genomics and proteomics will identify many more. Proteins are, however, costly to produce at the purity required for clinical use through current cell culture techniques, and it would be highly desirable to have available alternative methodologies. Secretion of proteins into milk and other biological fluids offers one such route to production of valuable proteins. For example, to bring on-stream a cell culture facility for drug production takes 4 to 5 years and costs over £100 million, whereas to produce a transgenic founder animal and develop a herd of cows derived from it would cost less than £5 million (Forsberg & Bishop 2002) and take no longer. The market for recombinant therapeutic proteins was US$12 billion in 1998, with an estimated 12% growth annually to 2006 (Jasuja 2000). For monoclonal antibodies, the market grew between 1999 and 2001 from $900 million to $3.5 billion, based on only 10 products, with a further 270 products under development.

For these reasons, several companies are in the process of developing products through either microinjection or nuclear transfer in sheep, goats, cattle and pigs. For secretion into milk the most commonly used promoters are derived from the β-lactoglobulin or α-S1-casein genes, and the products include α_2-antitrypsin, factor IX and fibrinogen. Other routes for secretion might include blood (human antibodies), urine and seminal plasma (growth hormone), each with different advantages and disadvantages: for example, one advantage of seminal plasma is the existence of the blood-testis barrier, which prevents compounds produced in the testis from acting systemically.

An alternative approach to the therapeutic use of transgenics is the modification of pig organs for transplantation into human patients. Apart from the dangers of viral transmission, this field has been slowed by the need to modify pig tissue antigens. The human immune system rejects pig tissue by an antibody response to the disaccharide galactose-α-1,3-galactose, which is present on the cell surface in porcine tissues but absent in humans. A great deal of effort is currently being put into the development of transgenic pigs lacking this antigenic epitope, and the availability of nuclear transfer in pigs should assist in their production.

The success of these procedures relies heavily on advancing requirement R3 for precision animal breeding, an understanding of gene expression and its contribution to phenotype.

Opportunities for this Century

We are at the threshold of an era where our assumptions of what traits can be addressed by breeding, how merit is assessed and the impact breeding may have will need to be completely revised, primarily due to developments in DNA technology. Although significant strides have been made in applying the principles of precision animal breeding to domestic livestock, there are good reasons to assume that the opportunities for livestock breeding will advance much further in this century. The size of the industry associated with livestock breeding will provide a pull for technical advances, many of which will also be applicable in other sectors.

One reason for optimism is the potential for implementing *genome-wide selection* (Meuwissen *et al.* 2001). Genome-wide selection will use the dense SNP maps emerging from genome sequencing projects. The philosophy of genome-wide selection differs fundamentally from the twentieth century approach of using DNA to enable selection to use marked quantitative trait loci (QTL) or individual genes: genome-wide selection is not concerned with how many QTL there actually are, or where they are located, but rather predicts BVs after weighing uncertainties associated with each small segment of DNA. Dauntingly expensive in the past, the new SNP technologies have reduced the costs of dense genotyping in each animal so that, if not affordable now, then it soon will be. Genome-wide selection tracks gene flow and segregation in all segments of the genome and avoids many of the problems associated with QTL discovery and usage within breeding schemes that has hampered the application of DNA technology to date.

To carry out genome-wide selection, breeders will need to obtain DNA from individuals, as is done now for pedigree testing. A large number of SNPs (tens of thousands) would then be typed in each sample. This vast array of genotypes for animals would be set alongside the recorded phenotypic data on performance, and subjected to an analysis that is complex and computer-intensive, but robust, and analogous to current genetic evaluations.

The potential benefits that genome-wide selection offers to livestock breeders are: (i) increased accuracy with minimal cost to inbreeding (Woolliams *et al.* 2002), (ii) ability to overcome age limitations for traits that can only be measured late in life, (iii) the opportunity to overcome or reduce sex limitations, or more generally limitations caused by measuring only special subsets, (iv) use in non-pedigree populations (e.g. identifying desirable DNA fragments in commercial populations that may be selected for within a nucleus), and (v) a direct link between the genetic evaluation and the genome.

In some sectors, additional costs might be directly offset through changes in the structure of the breeding programme. These benefits would be expected to drive further innovation in both the structure and goals of breeding schemes.

To exemplify the benefits offered by genome-wide selection in the context of the aims of precision animal breeding, a goal for turkey breeding involves the improvement of growth and the maintenance of egg production.

This goal is challenging owing to an antagonistic genetic correlation between these traits (Kranis *et al.* 2006); moreover, the genetic signal is stronger in growth than egg production, and the former is measurable in both the sexes, while egg production is measured only in mature females. For these reasons, traditional indices have a difficulty in making progress in growth without reducing egg production. The potential of genome-wide selection is that the genetic signal for egg production will be stronger in the young males because DNA fragments across the genome can be tracked directly from a turkey hen to her son, making it feasible to deliver the desired goals more precisely.

A further advance towards goal G1 will result from better predictive understanding of genetic correlations arising from fulfilling requirement R3. In particular, genetical genomics with its focus on understanding variation in gene expression will close a gap (the *genotype–phenotype gap*) in biology, linking variation in the genome to variation in flows through metabolic pathways, the fields of the geneticist and physiologist, respectively. Understanding the physiology mediating genetic advance will expand understanding of the physiological consequences. In turn, breeders might anticipate in what way, and to what degree, selection may have adverse consequences, and how these may be avoided.

It is therefore important to recognize that the paucity of data recorded on individual animals limits our horizons. The cliché: 'you can't manage what you do not measure' is directly relevant. Only through the accurate recording of performance and the analysis of records will the opportunities provided by these genetic advances be realized.

We need to make better use of data that are already available, by establishing monitoring procedures that generate usable databases, linking these to other databases containing genetic, management or other performance information, and ensuring public databases are open for analysis by scientists.

There are examples of good practice in this area such as the British Cattle Movement Service, which makes data available for specific projects. Barriers to such open availability of data need to be overcome where possible, most obviously in ensuring appropriate confidentiality and in promoting data quality when recording on farms where staff time for such activities is restricted.

High-throughput sensor technologies are also developing, mostly based on biospectroscopy, which has the potential to automate many aspects of recording, with increasing miniaturization and portability of the sensing equipment. This will include technology in milking machines for online hormone assays as indicators of fertility and disease, and for monitoring and recording in abattoirs.

Precision animal breeding has much to offer in disease control. Thirty-six farm animal diseases have a genetic basis for susceptibility (Bishop *et al.* 2003), but at present in the UK we do not monitor any of them in genetic databases. The cost of animal disease to the UK livestock industry is £1.7 billion annually, and the impact on production is 17% (between 35 and 50% in developing countries). Some (zoonotic) animal diseases are transmissible to humans.

One specific and far-reaching aspiration is to integrate genome-wide selection into routine disease surveillance within the UK. Currently, Defra surveillance provides little genetic information of value, yet given the potential of the approach to dispense with the need for pedigree information, genetic surveillance may be secured by taking DNA-containing samples from casualties paired to a control sample from another animal on the same farm. This will then directly lead to a genetic strategy, delivered through genome-wide selection, coupled to more rapid and direct gene discovery arising from bridging the genotype-to-phenotype gap that is intrinsic to the approach. While the opportunity is extremely promising, research is needed to justify and quantify the benefits in order to persuade decision makers of the value of this approach.

Models for Disease

Spontaneous mutations, particularly in rodents, have led to important advances in functional genomics. Examples are the Brattleboro rat, which provided a model for the study of vasopressin production and function, and *ob/ob* (leptin-deficient) and *db/db* (leptin receptor-deficient) mice, which led to the discovery of leptin and its role in obesity. Since the sequencing of the human and mouse genomes, and the opportunity to understand the roles of genes in disease, the use of mutants as tools for the discovery of gene function has become more widely used. At present, the principal tools used in these studies are strains of rodents (predominantly mice) with characterized genetic mutations. There are in general two methods of generating genetic mutations in rodents for the development of models for investigating gene function: by chemical mutagenesis or targeted transgenesis.

Chemical Mutagenesis

Many laboratories worldwide are generating random mutations in the mouse genome by treatment with chemical mutagens such as the alkylating agent *N*-ethyl-*N*-nitrosourea. This compound causes point mutations in the double-stranded DNA of spermatogonial stem cells through mismatching to an alkylated base. Treated mice have a 1000-fold increase in the rate of mutation. The mutations occur at random in the genomic DNA, and the offspring must be screened for abnormalities in the physiological system or organ of interest. Owing to their random nature, collaborative schemes have arisen to promulgate mutant strains between laboratories, so as to make efficient use of them. Examples of the application of these methods to particular systems are given for defects in the visual system by Thaung *et al.* (2002) and for cardiovascular disease by Svenson *et al.* (2003). At least 1000 strains of mice derived in this manner are currently in use for functional genomics studies.

Transgenesis and Gene Knockouts

Introduction of additional genetic material to the genome can be accomplished through transgenesis. This can be achieved either by pronuclear injection, where foreign DNA is injected into a fertilized egg, by viral transduction (Whitelaw *et al.* 2004) or by transfection into ES cells. Genetic markers can be used to provide for selection of transgenic cells, before transfer to an embryo using a procedure known as *gene trapping* (Skarnes *et al.* 1995).

Alternatively, where a gene sequence is available and a targeted deletion is required in order to study the gene function, genes can be selectively removed. This is achieved by homologous recombination with a mutated version of the gene in a synthetic construct, usually in mouse ES cells, but also potentially in somatic cells for nuclear transfer. ES cells are then introduced into a conceptus and adult mice derived from them, which are bred to produce offspring homozygous for the desired deletion. Co-transfected markers such as drug resistance genes can be used to allow selection of appropriate cells, and the deletion can be verified in offspring by Southern blotting. Heterozygotes and wild-type mice of the same strain are available as controls. The phenotype of the knockout progeny reveals the gene's normal role.

However, in some cases genetic models produced in mice are not closely applicable to human pathology, as gene functions differ between the species. As a result, other species are being investigated. For instance, the human cystic fibrosis mutation does not lead to a comparably severe condition in mice, and sheep may provide a more suitable model (Harris 1997). In addition, there is a project to create a pig model of neurodegenerative disease, culminating in the birth of pigs lacking functional copies of the ataxia-telangiectasia mutant (A-TM) gene. These pigs will provide an alternative animal model in which ataxia-telangiectasia (A-T) can be studied, which the current rodent models fail to reproduce. However, the techniques for producing such targeted mutations are time consuming and unpredictable, since the inability to derive ES cells that contribute to the germ line from domestic species necessitates the use of somatic nuclear transfer. An alternative to the mouse as an animal in which gene function can be studied is the zebrafish (*Danio rerio*). The advantages of this species are the large numbers of accessible, transparent embryos which can accelerate phenotype screening, and as a result the zebrafish is a productive model system for the genetic analysis of embryogenesis, organ development and related pathologies. Over 2000 mutations are currently available in more than 600 genes, and there is a highly developed international collaborative network in place for dissemination of these resources.

Opportunities for this Century

The deliberate production of animals with genetic disorders for research purposes clearly fails to meet goal G2 of precision animal breeding, and random mutagenesis with subsequent identification of effects on phenotype fails to meet G1, which calls for precision in predicting breeding outcomes. For these reasons, it is appropriate that breeding under these conditions should be subject to legislation controlling the use of animals for research; in the UK, this legislation is contained within the Animals (Scientific Procedures) Act 1986. Breeding animals with defined disorders for research purposes meets the objective of the refinement, reduction and replacement (3Rs) enshrined within the Act, since more precise models of genetic disease should lead to refinement and ultimately a reduction in the numbers of experiments performed, through generation of more accurate data.

An important advance in this field will be the isolation of ES cells from livestock species. This will permit the development of animal models for human disease in a way not currently possible. However, it should be noted that although research in this area has been underway for the past 20 years, and despite the generation of chimeras, to date germ line transmission of ES cell genomic information has not been achieved in livestock.

Clearly, there needs to be more research on the gene expression patterns characteristic of ES cells. While ES cells remain unavailable for livestock species and somatic cell nuclear transfer is relatively inefficient, enhancing the efficiency of targeted homologous recombination will be an important area for the future.

Conservation Breeding

Breeding animals for conservation differs from breeding for production purposes in three principal ways: the breeding aims are not focused on improving quantitative traits, but are based on population size and genetic variation; the numbers of individuals in endangered populations are by definition low; and stochastic environmental factors play a huge role in determining population viability. Conservation of endangered species and breeds has historically been the concern of two separate scientific communities, those involved in conservation of rare breeds of domestic animals and those working with wild animals.

Rare Breeds of Agricultural Livestock

About 40 species of domesticated mammals and birds are kept worldwide, although only six mammals (cattle, buffalo, sheep, goats, pigs and horses) and four birds (chickens, ducks, geese and turkeys) are widespread. Among these, approximately 7000 breeds have been produced since the beginning of domestication. In the UK, 70 breeds are recognized by the Rare Breeds Survival Trust (20 became extinct between 1900 and 1973, before the Trust was formed). The criteria for 'Rare Breed' status include a numerical basis, current trend in population size and a consideration of the period for which the population has been closed.

Efforts have been made to identify and prioritize breeds for conservation action as requiring either maintenance *in situ* or *ex situ*, or cryoconservation (a form of maintenance *ex situ*). Eding & Meuwissen (2001) explored a centrally organized approach to cryoconservation in gene banks, attempting to define objectively: (i) which breeds would or would not be stored, (ii) how many individuals to sample from each breed, and (iii) which individuals should be sampled so as to maximize the genetic variation that can be recreated from a gene bank. This method depended upon deriving relationships among breeds based upon DNA markers. Other related approaches use genetic distances (again based on markers) to define alternative measures of diversity rather than the concept of stored genetic variation that was used by Eding and Meuwissen. Weitzman (1992) provided rules for defining what may or may not be a legitimate utilitarian measure of diversity.

D'Arnoldi *et al.* (1998) and Caballero & Toro (2002) provide interesting discussions on these rules. In particular, Caballero & Toro (2002) point out that diversity in a species is concerned with diversity within breeds as well as between breeds, and that this aspect is neglected in methods based on genetic distance between breeds. However, as Woolliams (2004) points out, action is required based upon the best information available at the time, using the best practice that is feasible at the time and involving stakeholders at all stages.

3

Artificial Insemination and Cloning

INFERTILITIES IN HUMANS

There is a large number of reasons for developing infertilities in humans. By using IVF many more infertilities have been treated.

Male Sterility

In a fertile human the total count of sperms should be 15-20 million per ml. When the number of sperms decreases, the person is called oligospermic and the condition as oligospermia. Similarly, when there is very low number of motile sperms, the patient is called azoospermic and the phenomenon as azoospermia.

Generally in azoospermic humans the number of non-motile sperms is high. In addition, for IVF, the presence of motile sperms should be around 10-20,000 per ml.

Female Sterility

There are several reasons and types of female infertility; few of them are:

- Non-functional (inaccessible or absent) ovaries (some times ovaries of female are non-functional or inaccessible, therefore, healthy oocytes will not be produced. In such cases oocytes can be procured from donor for IVF),
- Non-functional (or absent) uterus (in this condition the oocytes are obtained from female and subjected to IVF, and embryo transferred to surrogate mother for pregnancy and further development),
- Idiopathic infertility (no definite cause is known for idiopathic infertility. Possibly it may be due to abnormal fertilization or failure in fertilization. Therefore, IVF followed by embryo transfer will cure this infertility),

- Tubal infertility (fallopian tubes receive oocytes from follicles of ovaries for fertilization. Any deformity in fallopian tubes results in tubal infertility. IVF followed by embryo transfer in uterus has replaced the function of fallopian tubes. Therefore, the patients will not be subjected to surgical treatment).

Who Benefits IVF

It is not likely that any patient of any age or any group of the society can go for treatment for IVF. There are several prerequisite conditions applicable for male and female partners.

The male should not be azoospermic; however, if he is azoospermic and desires for a baby, he can procure semen either from sperm bank or elsewhere. Besides, the female partner should also be healthy and fit for IVF because it is she who is to bear the pregnancy.

Therefore, the female must be such that:

- Any surgical procedure should not create any trouble to her,
- Her ovaries should be accessible for oocytes,
- Her uterus should accept and bear pregnancy for the required period of about nine months,
- Her cervical canal should be easily negotiable for embryo transfer.

How the Patients for IVF Treated

First of all the female aspirant for IVF is suggested to keep the records of her menstrual cycle upto six months. This information supports the doctors to take fruitful decision for the possibility of IVF. The female patient is hospitalized to ensure the date for collection of urine for the presence of luteinizing hormone (LH) surge, and to fix date for administrating human choripnic gonadotrophin (hCG). The hCG controls the final stage of formation of follicle and oocytes. The oocytes may be recovered during any menstrual cycles *i.e.* natural cycle, stimulated cycle or controlled cycle. According to the nature of three cycles, the treatment for IVF is recommended.

During natural cycle, spontaneous LH surge is found out at 3-6 h intervals by taking urine or plasma sample. Under the influence of LH surge fully developed follicles of both the ovaries burst and ova collected in fallopian tubes. In natural cycle, only one ova at a time is formed. Secondly, the stimulated cycle is created by administrating clomiphene and/or human menopausal gonadotrophin (hMG). Clomiphene (150 mg) per day is given for the 5th to 9th day of the cycle. However, the duration may be prolonged for 10 minutes but there will be more risk of uterus receptivity and abortion. The hMG alone or along with Clomiphene is also given that does not allow the high rate of abortion.

But several abnormalities are also caused by hMG. Consequently multiple follicles and oocytes, are stimulated instead of one. At this time hCG is also given to female patients so that inhibition of LH surge can be checked. The disadvantages associated with these are the ultrasonographic assessment of the hormonal effect and development of some of abnormal oocytes that result in abnormal pregnancy. Thirdly, hCG is administered at the optimum stage of maturation that arrests the development of follicle. However, it is difficult to predict the optimum time for hCG administration that varies for different preovulatory follicles. Patients need not stay in hospital for a long duration because the doctor can (to his own and patient's convenience) recover oocytes through laparoscopy. His success rate in oocyte recovery, embryo transfer and pregnancies are achieved by the stimulated and controlled cycles.

Indicators for Ovary Stimulation

It is very necessary to monitor the stimulation of ovary so that the oocytes can be collected at right time of development.

There are several parameters which are used as indicators for stimulation of ovary, these include:

- Prediction of LH surge on the basis of menstrual cycle,
- Rise in temperature of body during preovulatory and postovulatory days,
- Changes in cervical mucus as it is secreted five days before LH surge, goes maximum a day before LH surge and falls sharply thereafter,
- Estimation of progesterone,
- Levels of oestrogen in urine, etc.

Oocyte Recovery and Culture

Oocytes are recovered through recently developed equipment called laparoscopy. Using microscopic equipment the follicles can be observed and aspirated with the aspirating apparatus which consists of 23 cm long needle made up of stainless steel (with internal diameter of 1.6 mm and external diameter 2.1 mm) and 52 cm long tubing. The total space between needle and tubing is 1.2 ml. The ovulating follicles are selected for aspiration. It consists of ova of less than 1.5 cm. From such follicles fluid (4-12 ml) is aspirated by using needle and tubing. After aspiration wall of follicle gets collapsed. Fluid is of straw coloured some times it contains blood also. When there appears blood, heparinised culture medium should be added in tube so that blood clotting may be prevented. However, if fluid contains blood, ova could not be easily identified and isolated.

The aspirate is observed under the microscope for the presence of oocytes. If ova are not present in aspirate, the heparinised culture medium should be reflushed into follicles to get oocytes. The oocytes are identified under the microscope and incubated for 5-10 h depending on maturity of oocytes and follicles. The media used for oocyte culture and IVF are modified Ham's F10 medium, modified Whitten's medium, Earl's solution and Whittingham's T6 medium. Any one of these media may be used.

Semen Preparation

The semen is procured either (from husband of female partner or from semen bank or elsewhere. Husband's semen is collected at site when required through masturbation 60-90 minutes prior to insemination. Semen is subjected to centrifugation; sperm pellets are recovered and transferred to culture medium for suspending it again. It is then centrifuged and semen pellets are retransferred to culture medium. The sperms are incubated at 37°C for 30-60 minutes. The most active spermatozoa float at surface. Therefore, samples for IVF should be taken from surface of culture medium containing spermatozoa.

IN VITRO FERTILIZATION AND EMBRYO TRANSFER

Insemination is done by mixing about 1 ml culture medium containing oocytes with 1 ml of semen preparation (that contains 10,000-50,000 spermatozoa/ml). Oocytes are incubated for 12-13 h. Thereafter, they are observed keeping them in a Petri dish for the presence of two pronuclei and two polar bodies. Besides, the abnormal fertilized oocyte contains more than two pronuclei and granulation in cytoplasm.

First cleavage occurs after 24-30 h of insemination. Delay in cleavage after insemination confirms that the embryo is abnormal. Thus, the abnormal embryo is discarded; only normal embryo is used.

After fertilization, embryo should not be detained for el long time in vitro. It should be transferred to treated recipient female when it is in 2-4 cell stage. However, success has been achieved with transfer of 1-16 celled embryos. However, early embryo fails to survive in uterus. This stage is very critical, therefore, much care should be taken.

Some times, multiple embryo transfer is also done. It may increase the chance of embryo survival, but results in multiple pregnancy to which the female does not like. The transferred embryo stays in fallopian tube up to 8-16 cell stage, thereafter moves to uterus. Embryo of prolonged stage in vitro should not be transferred because of less chance of survival and pregnancy.

Embryo transfer should be done in an operation theatre. No anaesthesia should be given to female. She is given only Diazepam (Valium) of 10 mg orally before embryo transfer. The embryo is to be transferred through cervical canal; therefore, the patient is placed in lithotomy (*i.e.* in the position of knee and chest with tilting the head down) so that the fundus should be in lower position than cervical canal. The cervix is visualized by inserting a sterile bivalve vaginal speculum. Then uterine, cavity is aligned to cervical canal by manipulating the uterus.

Distance between fundus and external cervical is earlier measured through ultrasonography. The embryo is drawn into a teflon catheter along with 10 ml of tissue culture medium.

The catheter is gently inserted into culture medium. It should be made sure that the embryo has been transferred or not. The catheter is observed under microscope. Absence of embryo in catheter ensures for its transfer in uterus. Wrong placement of catheter or high amount of medium will result in tubal pregnancy or expulsion of embryo from uterine cavity. Therefore, care should be taken for proper handling. After embryo transfer, the female is allowed to take rest for about 2-7 h and abstain from intercourse for about 10 days. The patient follows the advice of doctor and undergoes for regular check up.

IN VITRO FERTILIZATION (IVF) TECHNOLOGY

The term in vitro means in glass or in artificial conditions, and IVF refers to the fact that fertilization of egg by sperm had occurred not in uterus but out side the uterus at artificially maintained optimum condition. In recent years the IVF technology has revolutionized the field of animal biotechnology because of production of more and more animals as compared to animal production through normal course.

For example, an animal produces about 4-5 offsprings in her life through normal reproduction, whereas through IVF technology the same can produce 50-80 offsprings in her life. Therefore, the IVF technology holds a great promise because a large number of animals may be produced and gene pool of animal population can also be improved. In India M.L. Madan, an animal embryo-biotechnologist at National Dairy Research Institute, Karnal (Haryana) has got success in producing more calves in cows. The IVF technology is very useful.

It involves the procedure:

- Taking out the eggs from ovaries of female donor,
- In vitro maturation of egg cultures kept in an incubator,
- Fertilization of the eggs in test tubes by semen obtained from superior male,
- Implantation of seven days old embryos in reproductive tract of other recipient female which acts as foster mother or surrogate mothers.

These are used only to serve as animal incubator and to deliver offsprings after normal gestation period. The surrogate mothers do not contribute any thing in terms of genetic make up since the same comes from the egg of donor mother and semen from artificial insemination.

In Vitro Maturation (IVM) of Oocytes

The immature oocytes are incubated in vitro so that they can be mature. However, immature oocytes should be taken out from follicles because they cannot mature in it but degenerate. Therefore, full potential of superovulation and all the oocytes can be utilized by IVF technology. Moreover, metabolic and hormonal requirement for oocytes during IVM should be found out so that the present rate of maturation (20%) could be improved. In majority of cases ovarian follicles never reach maturity and degenerate due to unknown causes. Possibly there may be genetic defects associated with them.

Culture of in Vitro Fertilized Embryos

IVF ofeggs is carried out in small droplets (microdroplets) of culture medium. Each microdroplet comprises of about 10 oocytes. The medium should be supplemented with penicillamine, hypotaurin, and epinephrine because they facilitate penetration of sperms into oocytes. Moreover, one dose of sperm is given that consists of about one million sperms per ml of medium. Thus, IVF embryo must be maintained at in vitro conditions for a few days so that it may develop into blastocyte.

It takes about seven days for sheep and goats and eight days for cattle. There are many laboratories where about 60 per cent IVF embryos of catties are cultured to blastocyte stage.

The term delivery from cultured embryo is very low due to occurrence of high loss during first two months of pregnancy. This may be due to abortion of foetuses arising from the presence of genetic defects. It should be noted that before birth about 80 per cent genes play a key role in differentiation and development of foetuses.

The oocytes which are forced to mature in vitro occasionally bears some defects. Some times environmental mutagenesis occurs in eggs, sperms or embryos.

Artificial culture media should be improved as oxygen may have toxic effect. Therefore, gas atmosphere should be carefully controlled. The other most successful method of IVM is to place the fertilized zygotes into agar (so that it may wrap around it) and implant them in oviduct of synchronized sheep or rabbit where the environment for early development of embryo is perfect. For early bovine embryo the oviduct of rabbit and sheep has been used as in vitro culture system.

MANIPULATION OF REPRODUCTION

Animals are a source of high protein, food, fibre and leather which men have been using since the time immemorial. In India milk and milk products have become a part of our civilization. Even in any small festival, demand of these two gets increased as several milk-based sweets and food items are prepared. Right from God worship to dining table use of milk and milk products is unavoidable. However, milk production could not keep pace with milk demand. Therefore, to meet milk demand synthetic milk and synthetic khoa are illegally sold in market even after several punishment threats declared by the Government. Due to unaware use of synthetic milk several infants have been reported to lose their lives

Production of food from animals is inexpensive and exhausting, and less efficient than plants. The most common farm hoofed animals are cattle, sheep, goats and pigs but unlike plants their management is expensive. Inspite of expenses rendered on nutrition, shelter and management, animals have been our companion since the dawn of civilization. Moreover, there is regional variation on availability of animal protein to human diet due to cost, culture and religion.

In biological sense, despite several similarities animals differ from plants in their reproduction systems such as number of gametes (ova), completion of life cycle, production of products, etc. Except pigs females of all common livestocks produce one egg at about monthly intervals.

Therefore, through biotechnological methods discovered in recent years, improvement in reproductive systems of animals, their milk proteins and milk-products, and several novel products which normally lack in animals can be carried out.

INSEMINATION

A male animal produces millions of sperms daily. Theoretically, it can inseminate females regularly and produce several offsprings. This excess capacity of male has been utilized through developing new technologies for artificial insemination which can be said as the first animal biotechnology.

The most effective factor that has increased the productivity of cattle is the artificial insemination. However, the breeder must replace the nature through artificial insemination if he ensures about the ovulation of female in herd together at a time. In contrast, if a breeder awaits for female to ovulate and then inseminate separately, the importance and economic significance of artificial insemination get reduced. Therefore, through artificial insemination technology increased number of females can be inseminated by a male.

Semen and Its Storage

In addition methods have been developed to produce semen from male by ejaculation. Semen ejaculate is collected and diluted (extended). Sperm motility and their number per milliliter are examined under the microscope. About 0.2 ml bull semen contains about 10 million motile sperms. The diluted sperms may be used fresh within a few days or cryopreserved at-196°C by using liquid nitrogen. The cryopreserved semen can be stored for a long time and easily transported across the states or countries. Thus, cryopreserved semen of a single male is capable of inseminating thousands of females of a country or other countries. For example, one ejaculate semen of a bull is sufficient to inseminate about 500 cows.

Ovulation Control

In many animals it is difficult to find out oestrous (sexual heat) in animals because it persists only for a few hours and occurs mostly at night. After ovulation (which is indicated by oestrous) females are inseminated. But in a herd it would be economical, easy and simplified management if females are inseminated at a times. However, it is possible only when all the female ovulate at a time; in practice it is not possible to get synchronization of oestrous. Moreover, it is possible to bring about ovulation in about 80 per cent of females by using hormones, for example progesteron and/or prostaglandin. These hormones regulate ovulation cycle of female and result in total synchrony of oestrous.

Sperm Sexing

Sperms are produced in the testes of males and ova in female's ovaries. Sperms and ova contain half of chromosomes as compared to somatic cells. An ovum possesses autosomes and one X chromosome. Similarly, a sperm contains autosomes and one Y chromosome. In animals sex is determined genetically *i.e.* by sex chromosomes. X chromosome determines femaleness and Y maleness. All the ova contain X chromosome, whereas a sperm consists of either X or Y chromosomes. One sperm ejaculate contains half X and half Y chromosomes.

In dairy industry demand of females is more than the males. Secondly, females have more desirable characteristics. The livestock industry prefers animals of one sex. Therefore, through artificial insemination technology X and Y chromosomes can be detected and sex of progenies determined accordingly.

There is a fluorescence dye (Hoechst 33342) that stains X and Y chromosomes with different intensities. Thus, these two chromosomes possibly can be separated by using an instrument, fluorescent activated cell sorter

(FACS). Sperms are present in the form of suspension. The FACS converts a suspension of sperms into microdroplets. Each droplet consists of a single sperm cell. Individual microdroplet passes through a laser beam. Microdroplets of different intensities are deflected into separate collection tubes as the fluorescence of dye is measured electronically. The sperms separated by using FACS have recently been used and pre-sexed calves have been produced through in vitro fertilization technique.

Moreover, FACS is very expensive and slow. It takes about 24 h to process one semen ejaculate, whereas the sperm cannot remain viable for a long time. Therefore, more refinement in technique is required for its use on a large scale.

Embryo Transfer

In 1890, the first case of developing pregnancy in rabbits through embryo transfer is known in literature. During 1930s the same method was used in goat and sheep. In catties cases of embryo transfer were reported after 1950 (BIOTOL series, 1992). Embryo transfer method cannot be used widely because of its high cost, technical difficulties and limited supply of embryo from superovulated donors. The embryo develops in foster mother (recipient) which simply acts as incubator and does not make any genetic contribution to the offspring. Secondly, ruminant female carries one pregnancy at a time as only one egg is produced and fertilized with male's sperm. Thus, there is a chance for increasing the number of egg production at a time and transfer of fertilized eggs *i.e.* embryo into uterus of less important foster mothers other than original female in farm animal.

Multiple Ovulation (Superovulation)

The reproductive cycle of ruminant female is such that the ovarian follicle of a non-pregnant female matures and releases single egg at a time. The time of ovulation differs in different animals, for example 21 days in cows and horse, 16 days in sheep and goats, etc. Normally ovulation occurs as a result of circulation of gonadotropic hormone. But by increasing the concentration of hormone the number of egg production gets increased. In well managed domestic catties 8-10 eggs are superovulated; the number may go to 60. However, this depends on health, nutrition, breed of animals and environment in which they live. Price (1991) has reviewed the current practice of multiple ovulation and superovulation. This technique has spread from cattle and sheep to goats, horses and deers.

He has emphasized that:

- General selection for increased litter size is useful only in sheep,
- The gonadotropic hormone induces superovulation in goats, sheeps and cattle but the response varies so much,

- Immunization against ovarian steroid hormones can increase litter size in sheep. Therefore, different molecular forms of follicle stimulation hormones (FSH) should be characterized in the farm animals.

Misra *et al.* have reported multiple ovulation and embryo transfer in Indian buffalo (Bubalus bubalis). After injecting the gonadotrophin the females are induced for superovulation. During follicular phase (second phase of oestrous cycle) about 20 ovarian follicles are induced. Eventually these grow and filled with fluid. The space of follicle which is filled with fluid is called antrum, and such follicles as antral follicles. In normal course, only one follicle develops and releases one egg after maturation. However, before ovulation the follicles laying against surface of ovary looks large sized (8 mm in sheep and pigs, 15 mm in cattle). Therefore, immature oocytes from follicles of donor females are recovered surgically by using laparoscope. The females to be superovulated are frequently injected with prostaglandin F2a (PGF2a) so that synchronized oestrous could develop in them. After 10 days of oestrous they are injected with hormone FSH up to 4 days followed by PGF2a treatment so that oestrous may be maintained. The FSH treatment induces superovulation. The females are artificially inseminated. The eggs are fertilized. After fertilization embryos undergo developmental stages.

Multiple Ovulation with Embryo Transfer (MOET)

After 6-8 days of fertilization (for sheep and goat) the embryos are recovered. During this stage embryos have come to morula or blastula atage and remain in female's oviduct. In cattle the embryos are recovered without surgery by inserting a catheter into oviduct. A saline solution is drained in oviduct and embryos flushed out through catheter into a storage bottle. The oviduct of sheep is small, therefore, it is exposed surgically and embryos are recovered by syringe adopting the same method for cattle. In storage bottle embryos settle down and solution decanted. Embryos with small volume of saline are poured into a Petri dish. Then they are examined under microscope. The identified embryos are transferred into synchronized recipient females (l as artificial insemination), when they are in 6-8 days of cycle of embryo development, stored and transported or manipulated as desired. In general, one superovulated female results into 5-6 progenies. In cattle 50-60 per cent of pregnancy can be achieved from embryo transfer method.

Embryo Splitting (Demi-Embryo)

Embryo of blastocyst stage is differentiated into two regions, trophoectoderm and inner cell mass (ICM). Trophoectoderm is a single layer of trophoblast cells lining the inner side of zone surrounding the blastoceal (a fluid filled cavity). Zona becomes the foetal membrane bf placenta. The ICM is the mass of cells which develops into foetus. The embryos of blastocyst stage can be split into equal two halves and transferred to females to produce identical twins. Thus, embryo splitting technology has increased the rate of pregnancy. To carry out embryo splitting, the blastocyst stage embryos are transferred for a few minutes into a cell culture medium consisting of hypertonic sucrose and bovine serum albumin.

The medium of high osmotic strength enters into cell membrane of the embryo (zona pellucida) and cells contract due to exosmosis. The bovine albumin serum attaches to zona pellucida and provides negative charge to the embryo. Then it is transferred into a plastic Petri dish containing standard cell culture medium. The outer membrane of embryo is negatively charged and Petri dish positively. Embryo sticks to the surface of Petri dish due to development of electrostatic interaction of two charges. The Petri dish is kept on stage of an inverted microscope. A micromanipulator equipped with fine surgical blade is used to cut the embryo roughly into two halves with minimal damage to the cells. The hemispherical mass of half embryo, demi-embryo or semi-embryo reforms spheres.

However, it is not necessary for successful implantation of demi-embryo that it should be enclosed in a zona pellucida. The demi-embryos are transferred into oviduct of synchronized recipients as described for normal embryo transfer. At this stage it is very necessary to know the situation for successful embryo implantation that the wall of uterus of synchronized female recipient is waiting for embryo, and embryo is prepared to send chemical signals from the trophoblast.

To increase the rate of pregnancy, the embryo can be cut into equal three or four pieces, and transplanted in synchronized females. But too small trophoblast will not induce the uterus for pregnancy. So far it is unknown how cells of ICM are required for successful pregnancy, but it is clear that increasing in embryo splitting will have less probability of pregnancy.

Embryo biopsy (removal of small number of cells for genetic analysis) should be combined with splitting so that the twins which will be produced should be identical and of known genotype. Biopsy is very necessary in breeding so that sex and genetic diseases could be detected. In case the embryo has any genetic disease, it can be prevented from implantation in recipient females.

Embryo Sexing

Before the implantation of embryo its sex should be detected from the biopsy sample. The principle for sexing is very common. The presence of Y chromosome makes the offsprings male and that of X makes female. The second method is the use of polymerase chain reaction (PCR) machine in sex detection. PCR amplifies DNA sequence of Y chromosome and reaction products can be seen directly. Handy side *et al.* isolated single blastomere from early embryo from a womb, amplified DNA sequences of Y chromosome and carried out embryo sexing before implantation into uterus. It is true that the PCR was first commercially implemented for embryo sexing of livestock. Now the sexed embryos are available commercially but these are very costly.

EMBRYO CLONING

A clone is a population of cells or organisms derived asexually from a single ancestor. They are genetically identical to each other to their common ancestor. Cloning means the production of exact genetic replica copies of an individual.

They can not be considered as an offspring but simply the copy of a given individual. Much work has been done on cloning in plants and microorganisms. However, the techniques used in plants can not be applied for animals. Moreover, many animals from a single genetically superior embryo can be produced. Still there is no method of finding out which embryos are capable of cloning. It is useless to clone an embryo if it is not superior.

Quadriparental Hybrids

During 1960s, Beatrice Mintz at Cancer Research Institute, Philadelphia (USA) demonstrated the interesting experimentation. She carried out fusion of embryos of two different species of mouse. This resulted in the formation of a single embryo which finally developed into a normal healthy animal having four parents.

Embryo A was derived from the cross of male x female of one species, and embryo B derived after cross of male x female of the other species. The embryo A and B were united together and produced a single mass. In this experiment Mintz removed zona pellucida membrane of two early embryos and placed them in a suitable culture medium.

The embryonic cells of the two embryos of blastula stage united randomly into a single mass of double sized embryo (blastocyst). A fresh membrane developed around the embryo. Then the embryo was transferred into the uterus of a foster mother. The foster mother was mated with a sterile male to bring her into proper stage for implantation. The first offspring with four parents was born in 1965. Similarly, exciting experiments have been done on another mammal. Following the same technique a hybrid of goat and sheep named geep was produced.

At present two types of techniques for embryo cloning *viz.*, nuclear transplantation (transfer) and embryonic stem cells, are being developed.

Nuclear Transplantation

Nuclear transplantation (also nuclear transfer) involves removal of a single blastomere from a cleavage stage embryo with a fine micropipette of glass, and placing it under the outer membrane of an unfertilized mature enucleated oocyte (whose haploid nucleus has been removed by using micropipette or destroyed by UV light). For the first time in 1955, Robert Briggs and Tom King at Cancer Research Institute, Philadelphia (USA) carried out nuclear transplantation experiment on embryonic cells of frog. They transferred nucleus of undifferentiated blastula (a stage soon after fertilization of egg) into an enucleated egg cell. They noticed the normal development of the embryo.

When they performed serial transplantation of differentiated nucleus from late gastrula (a stage after blastula) into a nucleus-free unfertilized egg, abnormal embryos were formed. This shows that cell nucleus is differentiated with embryo development. In 1960s, J.B. Gurdon at Oxford University, U.K. transferred differentiated intestinal nucleus of a frog into nucleus-free unfertilized egg of different amphibian species (Xenopus laevis). The embryo developed into tadpole and matured into frog. This new enucleated cell developed into normal embryo. Any damage to the donor nucleus during transplantation leads to abnormal development.

DOLLY-The First Mammalian Clone

'Dolly', the worlds' first mammalian clone has been created from a fully differentiated non-reproductive cell of an adult sheep. It was born in February, 1996. The name Dolly has been given after an American country singer, Dolly Parton. In 1995, Ian Wilmut and his team of researchers at Roslin Institute, Edinburgh, Scotland, took udder (a fully differentiated tissue) from six year old sheep, Fin Dorset Ewe, and placed it in special solution that controlled cell cycle of cell division.

The cell was deprived off certain nutrient. At the same time an unfertilized egg was obtained from another adult sheep. Its nucleus was carefully removed leaving the intact cytoplasm in egg. The nucleus of udder cell was taken out and transferred into nucleus-free egg. This was facilitated by applying mild electric sock. The newly transplanted nucleus soon became functional according to the new cytoplasm in which it had been artificially transferred. This viable combination underwent cleavage like normal zygote. This so called embryo was then transplanted into the uterus of a third adult sheep (surrogate mother/ foster mother) for its further development.

Finally, a normal healthy little lamb, Dolly was born in February, 1996 which was genetically similar to the clone mother from which nuclear DNA was taken out. It does not have any similarity with that sheep from which egg was taken out or surrogate mother because they did not contribute any genetic character. Thus, Dolly has only a single parent because she has born asexually, a characteristic feature found in lower forms of animal life, not in mammals.

Although behind this great success the rate of success is very slow, yet it has given some hope to embryo-biotechnologists to bring about refinement. Out of 277 nuclei transferred singly to enucleated egg, only 29 eggs grew into embryos. Out of these, only 13 embryo could be successfully transplanted into surrogate mothers. Of these only one ewe was successful in giving birth to an offspring, Dolly.

The significant considerations that can be derived from this experiment are that:

- The genes of differentiated cells have inherent totipotency,
- The interplay between the regulatory system of the genome in nucleus and the cytoplasmic factors of egg may make a cell totipotent,
- Possibly the cytoplasm of enucleated egg makes the transplanted nucleus totipotent like that of normal fertilized egg nucleus,
- The maternally derived information in egg cytoplasm has an important role in cleavage which usually occurs after fertilization.

Hence, it is the egg cytoplasm but not the nucleus that regulates cleavage. Because the udder cell nucleus has limited potential for mitosis; it is the egg cytoplasm that interacted intracellularly with udder cell nucleus and stimulated to undergo repeated mitosis,

- Carbon copy of the adult sheep could be produced without involving sperms from male partner,
- The cloned animal produced via nuclear transplan-tation technique will be capable of restoring fertility as in 1998 Dolly gave birth to a little lamb named Bonny.

Embryonic Stem (ES) Cells

Cloning of mice could not be done as in sheep via nuclear transplantation. This was due to acceleration of developmental programmes of embryo. However, it is evident that before first embryonic division the cell has started its process of differentiation.

Therefore, for cloning of mice an alternative approach has been made *i.e.* the use of ES cells. A blastocyst of mouse is place in culture condition. The inner cells that form future foetus continue to divide and remain in undifferentiated totipotent state as ES cells. There is a peptide growth factor known as leukaemia inhibitory factor (LIF) which establishes and maintains ES cell lines.

The ES cell lines will be very useful in the area of production of transgenic animals. However, the ES cells are used in two different ways: a small number of ES cells can be injected into blastocoel space of a blastocyst. The ES cells get mixed with inner mass of cells of blastocyst to produce a chimera mouse which is a mixture of two cell genotype having the patches of different coloured fur. Crossing of male and female chimera will allow selection of homozygous mice derived from ES cells.

In Vitro Fertilization and Embryo Transfer in Humans

Due to several complicacies associated with humans, there has been the birth of in vitro fertilization technology. The IVF technology is a boon to childless couples. Initially this technique was pioneered by Prof. Robert Winston during 1970s which was later applied by P. Steptoe and R. Edwards for the production of first test tube baby, Louse Joy Brown, who was born on July 25, 1978. Since then more than 25,000 test tube babies have been produced so far through out the world.

4

Statistical Methods in Animal Breeding

INTRODUCTION

A recent GIFT workshop had two papers that discussed future dairy cattle research. The two papers were in good agreement of the future statistical needs. These included methods for test day models, international comparisons, non additive variance, non-linear models and individual gene models. They also highlighted a trend to more sophisticated analysis leading to less biased predictions and more progress at the expense of greater variance or risk. Cost of analysis was suggested to be small compared to the cost of collection of data. There was a concern that uncertainties in parameters might erode possible gains. There was also a hope that prediction were robust to bad luck. Variance parameter estimation plays an integral role in several of these topics. We therefore intend to review this area hoping to identify themes that will lead to more rapid change. Frem Narain has made an outstanding contribution to statistical genetics, including the application of statistics to plant and animal breeding, so we think our small paper contribution is particularly apt to this volume in honour of Prem's 70th birthday

USING PLI TO BREED

In the last 10 years, genetic indices for dairy cattle across the world have been moving toward the breeding of a more sustainable cow with a shift towards lifetime production rather than lactation production. Consequently, the UK, has moved from Profit Index (£PIN) which was totally production based to Profitable Lifetime Index (£PLI) which also includes non-productive traits such as fertility, lifespan and somatic cell count.

Genetic trends in dairy cows have been for increasing milk production but this has led to declining genetic merit for butterfat and protein per cent, fertility and lifespan. Infertility is estimated to cost the average 100-cow herd around £22,000 per year and is the primary reason for dairy cow culling.

However, when selecting AI sires, many farmers do not take account of sire fertility index, despite the fact that DairyCo has published it with sire proofs for over five years now.

This is the message from Marco Winters, head of genetics for DairyCo Breeding+ who highlights that: "There is a need for farmers to gain a better understanding of breeding and sire selection so that a clear breeding plan can be established."

"As farmers have used bulls with higher and higher Predicted Transmitting Abilities (PTAs) for milk production, their PTAs for calving interval have got worse. In other words, they've got longer," says Mr Winters.

"However, since 2005 when the Fertilisty Index was first launched, selection continued to favour milk production but the genetics for calving interval started to improve. This is excellent news, because it tells us that it's possible to continue breeding for better production without sacrificing fertility and offers the realistic hope that future improvements can be made through breeding."

To investigate the practical impact of sire fertility index on daughter fertility, Dr David Mackey of CAFRE has retrospectively analysed the fertility records of Greenmount's Future Herd and several other pedigree and non-pedigree herds throughout Northern Ireland.

"While it takes data from a lot of cows to demonstrate genetic trends, the benefits of selecting sires with a positive fertility index can be clearly seen through both a reduced calving interval and an increase in conception rate."

Dr Mackey goes on to comment that: "Had we know that some of the sires used before 2005 carried a negative Fertility Index, they would not have been selected. Since then only bulls with a positive Fertility Index are selected and we should start seeing the benefits of this in the next few years." A number of dairy farmers in Northern Ireland actively use £PLI in their sire selection each year and see the benefits of it, including Drew McConnell from Newtonstewart, Gary McHenry from Lurgan and Kenneth Montgomery from Eglinton.

Drew comments that he "Always selects bulls on £PLI, ideally those with a £PLI of £180 or more," but goes on to say that "they need to have positive milk figures and positive fertility as well."

Analysis of Drew's breeding records established that, on average, each one-point increase in a sires Fertility Index reduced the average calving interval of his daughters by almost a day, twice that estimated by DairyCo (0.5 days).

This demonstrates the benefit of including positive fertility in your sire selection criteria. Drew also says that "I cull a lot of cows for mastitis and I'm trying to address this through breeding by selecting bulls with a negative SCC score."

Similarily, Gary McHenry focuses on £PLI as a simple and effective screening tool.

"As we already have the genetics to produce 9,200 litres, our focus isn't so much on milk yield but on protein percentage, fertility and somatic cell count.

"We only look at bulls with a PLI of £150 or more, and then select bulls with protein percentage above +0.10 per cent, fertility index above five and SCC less than-10. These traits are all heritable and will contribute to higher profitability in future."

Mr Winters comments: "Farmers like Drew and Gary who are paying attention to and adopted the strategy of selecting on fertility are to be congratulated, and the AI companies have played an important role in the process through the increased listing of positive fertility bulls in their catalogues."

However, fertility isn't the only non-productive trait considered within the £PLI index. Bulls also carry proofs for Lifespan and Somatic Cell Count, and these should also be considered when selecting bulls.

Mr Winters suggests that "Sires should be screened on £PLI in the first instance, and then check the fitness traits important in your herd and ensure you are using bulls that have strengths in the right areas. For many herds, the area which will need particular improvement is fertility, and today, it's possible to find more bulls on the market than ever before which will transmit good fertility on to their daughters."

Advancing Genomic Selection

Genomic selection supplements the traditional method of genetic evaluation with information on the DNA of an animal to improve the accuracy of identifying genetically elite animals. Noirin McHugh and Donagh Berry from Teagasc say that greater gains can be achieved by applying genomic selection to females as well as males.

Uptake of Genomic Selection in Ireland

Since its initial launch in 2009 the uptake of genomic selection in Ireland has been strong. In 2009 genomically selected (GS) bulls accounted for 34 per cent of the total AI straws sold in Ireland; 2010 saw this figure increase further to 40 per cent. One of the main reasons for the rapid adoption of this technology by Irish dairy farmers is the greater EBI of the GS bulls (218 in 2010) compared to the Irish daughter proven bulls (€146 in 2010). The reliability of available young bulls increased from 32 per cent prior to genomic selection to 54 per cent with genomic selection. Although the reliability of the GS bulls is lower than for most Irish proven bulls, the associated increase in the ebI of the Gs bulls has seen a strong uptake among Irish farmers. However, as the reliability of the GS bulls remains relatively low, farmers are adhering to the advice given and using an average of four Gs bulls per herd to spread their risk.

Advances In Genomic Selection in Ireland

Fundamental to generating a benefit from genomic selection, is knowledge on the DNA profile most suited to Ireland. A population of Irish proven animals, commonly known as a training population, is used to relate the DNA profile of the individual animals to their genetic merit for the array of traits evaluated by the ICBF.

Through international collaboration, the training population in Ireland has more than quadrupled from 945 domestically proven animals in 2009 to 4,196 in 2011. This training population size is still lower than in some countries and is reflected in lower increases in reliabilities from genomic selection. When genomic selection was initially launched in 2009 the service was only available for animals that were at least 50 per cent Holstein due to the small size of the training population for Friesians.

Genomic selection, however, is now available for Friesians. other recent advances include the development of genomic selection for linear type traits; initial results show an increase in the reliability of the type traits by approximately 10 percentage units.

Genomic Selection on Farm

Genomic selection is now available to all dairy farmers. ICBF provide two genotyping services: the low density "3k" (i.e. 2,900 markers) platform or the normal density "50k" (54,001 markers) panel. because animals inherit large chunks of their dna from their parents, animals that have both their sire and maternal grandsire genotypes in the ICBF database (now 9,000 genotypes) can avail of the lower-cost 3k genotype service; while other animals can avail of the 50k genotype service; the cost of each service is €50 and €99, respectively. Farmers can log onto the ICBF website and see which animals can avail of the lower cost service.

The Use of Females To Exploit Genomic Selection

To date the focus of genomic selection has been on the male side; however, the female has an increasingly important role to play. Studies have shown that the genotyping of females can substantially increase the size of the training population thereby improving the accuracy of the genomic ebIs. The genotyping of females can potentially result in a three fold increase in the rate of genetic gain. An economic analysis con irms that genotyping females can result in very high rates of return on the investment.

Although genomic selection has progressed rapidly since its inception, there are still many other areas where further gains can be made. Our continued role in international collaborations ensures that Ireland remains at the forefront of this cutting edge technology.

The swapping of genotypes with international partners ensures that the training population will continue to grow thereby increasing the reliability of our genomic EBIs. Access to more DNA markers may increase genetic gain further and expedite its application to multiple breeds.

The success of genomic selection has been due to the support and involvement of the entire industry, from research carried out by Teagasc, its implementation by the ICbF, the investment in infrastructure at Weatherbys to generate the DNA Profile, and the large uptake by dairy farmers. Teagasc and the ICBF will continue to work closely with the industry to further enhance genomic selection and ensure that the benefits are realised at farm level.

PEERING AT GENES TO DETECT ORIGIN OF CATTLE DISEASES

The eyes are said to be the windows into the soul of humans, but in the case of cattle, they may hold clues to overall animal health. A closer look at pinkeye is offering insight into other costly bacterial diseases as well.

Researchers at the Agricultural Research Service's Roman L. Hruska US Meat Animal Research Centre (USMARC) in Clay Centre, Nebraska, have found that genes are linked to the incidence of several diseases.

Eduardo Casas, a geneticist in the Genetics and Breeding Research Unit at the time of the study, and former ARS scientist Gary Snowder discovered a quantitative trait locus (QTL), or location, on bovine chromosome 20 that is associated with pinkeye, foot rot, and bovine respiratory disease.

Cow receiving treatment for pinkeye at the U.S. Meat Animal Research Centre, Clay Centre, Nebraska.

Looking Beyond the Pathogens

Scientists have known for some time what causes pinkeye, also known as "infectious bovine keratoconjunctivitis," and other diseases. The bacteria Moraxella bovis is the most common pathogen associated with pinkeye.

Pathogens associated with bovine respiratory disease include viruses, such as bovine viral diarrhea virus, bacteria, like Mannheimia hemolitica, and mycoplasma. Fusobacterium necrophorum and Porphyromonas spp. are the main bacterial pathogens for foot rot, or infectious pododermatitis.

"Scientists have spent a lot of effort and money studying the pathogens that make animals sick," says Mr Casas, who is now research leader for ARS's Ruminant Diseases and Immunology Research Unit in Ames, Iowa.

"We've made a lot of progress, but the microbes are still around. Therefore, the main focus of this research was to look at diseases from the animal's point of view."

Mr Casas' approach was to examine the genetic makeup of cattle for evidence of genes associated with conferring resistance or tolerance to diseases. His initial study focused only on pinkeye because it's easy to see and measure in cattle, he says.

Different breeds vary in their pinkeye tolerance. For example, Herefords are very susceptible, but Brahmans are highly resistant. With this in mind, a Brahman-Hereford crossbreed sire was mated to other breeds to yield more than 540 offspring.

USMARC, which has more than 6,000 head of cattle, provided an ideal location to study different breeds affected by pathogenic diseases. (K7648-3)

"This particular bull was heterozygous for all genes that would confer tolerance to pinkeye," Mr Casas says.

"Half of the offspring inherited the resistant gene, and the other half inherited the susceptible gene."

When scientists looked at 36 offspring affected by pinkeye, they found that regions on chromosomes one and 20 harbored genes that influence the presence of bacteria, but no strong linkage to a QTL was identified. So the team took a different approach.

A Tale of Three

Following up on a theory that the immune system is influenced by various genes, Mr Casas and Mr Snowder conducted a second study. They combined the incidences of three highly prevalent bacterial diseases affecting feedlot cattle—pinkeye, foot rot, and bovine respiratory disease.

"When you put all three diseases together, you're looking at the overall health of the animal, or resistance to multiple diseases, rather than a disease-specific response," Mr Snowder says.

"In other words, the particular loci affecting an individual disease may not be easy to pick up, but it might be easier to pick up markers that are related to the general health of the animal."

Selection for disease resistance is one of several possible interventions to prevent or reduce economic loss associated with animal disease and to improve animal welfare, according to Mr Casas and Mr Snowder.

A common condition affecting breeding-age beef heifers, pinkeye has a marked economic impact on the cattle industry—costing an estimated $150 million a year due to lower weight gains, decreased milk production, and treatment. Although not fatal, this highly contagious disease can affect up to 80 per cent of a herd. Calves being weaned are even more susceptible and can lose as much as 10 per cent of their body fat if they contract the disease. Bovine respiratory disease—pneumonia—is the most common and costly feedlot disease in the US. It accounts for 75 per cent of feedlot morbidity and up to 70 per cent of all deaths. Economic losses to cattle producers exceed $1 billion annually from animal deaths, reduced weight gain, lower feed efficiency, treatment costs, and poor-quality meat and hide products.

While foot rot is not as expensive as other diseases, it is estimated to cost dairy producers about $120 to $350 per animal. Foot rot causes lameness and leads to reduced milk yields, lower reproduction performance, increased involuntary cull rates, and discarded milk.

Producers have been managing these diseases with various treatments and management practices. USMARC, which has more than 6,000 head of cattle, provided an ideal location to study different breeds affected by pathogenic diseases.

Geneticist Eduardo Casas (right) and technician Sandra Nejezchleb prepare bovine DNA samples for mass-spectrometry. Results are used to assess whether changes in DNA sequence are associated with pinkeye, also known as "infectious bovine keratoconjunctivitis." (D2236-9)

Breeds Apart

In addition to the Brahman-Hereford family studied in the pinkeye experiment, three other half-sibling families were produced to detect QTLs associated with combined incidences of the three diseases.

The second half-sibling family was developed from a Brahman-Angus sire and produced 176 offspring. A Piedmontese-Angus sire fathered 209 calves, and a Belgian Blue-MARC III (part Red Poll, Pinzgauer, Hereford, and Angus) sire produced 246 offspring.

Researchers used microsatellite markers—short, repetitive DNA sequences used as genetic markers to track inheritance—to screen the genome of each family. Informative markers were chosen within a family based on their location in each chromosome.

All animals were observed daily throughout their life-span for pinkeye, pneumonia, and foot rot and treated when symptoms occurred. The 240 calves infected by one or more of the diseases were classified as affected by a microbial pathogenic disease and coded. Analysis of DNA blood samples taken from these animals revealed QTLs for disease activity.

Though scientists have discovered genetic locations that may influence resistance or susceptibility to bacterial diseases, there's more to do. "We don't know what the gene or genes are yet, and that's what we are working on," Mr Casas says. More study needs to be done to confirm the association between the genes and disease."

"What's interesting about the markers on chromosome 20 is that they are in very close proximity to other markers related to other diseases. That particular region may have a significant effect on the general health of animals," Mr Casas says.

Additional studies are under way to detect genes associated with reduced susceptibility to bacterial diseases, including Johne's disease and bovine viral diarrhea.

"The costs for treating animals that have these diseases are enormous," Mr Casas says.

"Identifying genes responsible would provide an opportunity for effective crossbreeding to produce animals with increased disease tolerance, which would greatly reduce the economic impact to the cattle industry."

BOOSTING BREED PLAN

Three funding terms of Beef CRC have ensured that new, hard-to-measure but economically very important traits have been added to BREEDPLAN, significantly increasing BREEDPLAN's value to the beef industry. Prediction equations based on DNA information from the new high density panels will be delivered to BREEDPLAN

Since the Beef CRC began contributing data to BREEDPLAN in 1993, performance-recorded Australian cattle breeds have shown steadily rising rates of genetic gain. Now a wave of new genetic information is coming through, made possible by the inclusion of new DNA-based genomic technologies.

Breedplan is built on "phenotypic" data, which comprises measured records of economically important characteristics of an animal. About a decade ago, technologies began appearing that seemed to promise to greatly reduce the enormous labour and cost of phenotypic recording by identifying the genes associated with these economically important traits.

At the time, "genotypic" (DNA-based) data promised a revolution that would leave the old methods in the dust. However, according to one of BREEDPLAN's founders, Dr Hans Graser, Director of the Animal Genetics and Breeding Unit (AGBU), the process has proved not to be so simple. There is now a clear realisation that gene interactions are far more complicated than was originally thought.

Forecasts of dramatic advances through genomics were made on the assumption that a handful of genes would explain most of the genetic variation for a trait.

However once the Bovine Genome was sequenced in 2006 and the full assembly released in 2009, it became clear that things were not as straightforward as expected. It turns out,Dr Graser says, that hundreds, even thousands of interacting genes may be at work to control a single trait.

Technology has produced some answers to this complexity, including the single-nucleotide polymorphism (SNP, pronounced "snip") chip. These chips allow tens or hundreds of thousands of genetic sequence differences to be quickly compared against each other, revolutionising the business of gene discovery.

The Beef CRC has used this technology to record genetic differences from thousands of animals, even reaching back into its archival material to genotype long-dead cattle used in Beef CRC's first term. A few CRC animals have also been fully sequenced, a procedure that was prohibitively expensive only a few years ago, to contribute to the global knowledge of beef cattle genomics.

That information now sits in the Beef CRC databases at AGBU.

"The genomics database will contribute to BREEDPLAN when DNA-based prediction equations are accurate enough to provide useful outcomes for industry," Dr Graser says, noting that this is a Beef CRC objective for mid-2012.

"Early on, we thought we could use genomics to explain 50 per cent of the genetic variation which is the equivalent of an EBV with an accuracy of 70 per cent," Dr Graser says. "Now the aim is to explain 15 per cent of variation, which equates to an accuracy of 40 per cent. To get higher accuracies, we need more recorded animals."

"For most traits that are phenotypically recorded, we get accuracies that are higher than we are currently achieving with DNA markers 40-60 per cent for the animals we have on record."

Breedplan will use both sources of information, with genomic data providing an independent cross reference for the phenotypic information. Given the current pace of progress, Dr Graser expects that by mid-2012, Beef CRC researchers will have developed improved equations that are strong enough to provide a useful prediction value for traits determined through use of DNA.

However, he warns the value of the data will be compromised unless researchers are able to validate the prediction equations against performance in animals that are completely independent of animals in the discovery populations.

"We are coming back full circle," Dr Graser said. "At the start of the CRC's second term, it was the genotyping that was expensive. Ten years later, genotyping costs have dropped dramatically while phenotyping costs have increased due to higherlabor costs and economies of scale have gone down."

The Angus, Hereford, Charolais, Limousin and Brahman societies have established Breeding Information Nucleus (BINs) herds which aim to amass phenotypic data. DNA samples from every BIN calf are being col-lected, because, Dr Graser said, the genomics revolution is only just gathering pace: the most promising developments are yet to come.

The Beef CRC is currently undertaking a bid for a five-year extension that will allow it to wrap up any outstanding research and industry delivery threads. If successful, CRC researchers will utilise the BIN herds to underpin the ongoing research activities.

Given the extremely rapid changes in genomic technologies over the past five years, Dr Graser predicts that over the next five years, DNA sequencing may become so cheap and fast that the existing SNP chips may be made redundant because the full sequence will be readily available for all animals.

If the phenotypic information is available to back up the DNA data, then it is likely that at that time, genomics will begin to exponentially accelerate genetic gain, as many had hoped it might have done by now. However, the greatest requirement to achieve the desired levels of genetic gain is accurate measurements on many more animals than are currently available.

New Traits Added to BREEDPLAN Through Beef CRC Research

The introduction of scanned carcase traits to the breeding program as a result of the BREEDPLAN validation program and Beef CRC phase one, increased the dollar profit per cow joined relative to industry recording of weights alone. The subsequent inclusion of reproduction traits and actual carcase and feed efficiency traits based on the first and second phases of Beef CRC increased genetic progress even further. As at 2008, expected profit per cow meeting this market specification was around \$90 and this was increasing at over five dollars/year.

The difference between the 2008 profit value and the profit value that would have occurred in 2008 if only growth rates were available (around \$60 by extrapolation) is the incremental value of the new traits added into BREEDPLAN due to CRC research.

Selecting Beef Animals for Feed Efficiency Improves Carcase Traits

Selecting beef animals for improved feed efficiency will also result in larger and leaner carcases, according to Teagasc.

Selection for improved feed efficiency will not have any unfavourable repercussions for carcass traits. And, if anything, selection for improved feed efficiency – lower food conversion ratio (FCR) and lower residual food intake (RFI) – will result in larger, leaner carcasses with better conformation.

Some good news for beef producers from Teagasc's John Crowley, who led a study to quantify the genetic relationship between feed efficiency and carcass fat (CF), carcass conformation (CC) and carcass weight (CW) in Irish beef cattle.

"Feed is the largest variable cost on farms and there is much interest in selecting for improved feed efficiency. But prior to recommending selection on any trait, the expected response to selection on other economically important traits needs to be quantified," he explained to delegates at this year's British Society of Animal Science's annual conference, held at Queen's University, Belfast.

Data for this analysis originated from two separate sources. The first data set consisted of feed intake (FI) and bodyweight (BW) records collected at the Irish bull performance test station. Different measures of feed efficiency (FE) were subsequently calculated.

The second dataset consisted of information on CC, CF and CW on 2,566,969 animals slaughtered in 30 abattoirs in Ireland during 2005, 2006, 2007 and 2008. In the present study, the EUROP classification grades were transformed to a 15-point linear scale with a carcass fat score of 1 implying low fat coverage and a carcass conformation score of 1 implying poor conformation.

Only males slaughtered between 300 and 1,200 days of age and females slaughtered between 300 and 875 days of age were retained. And only animals with Aberdeen Angus (AA), Belgian Blue (BB), Charolais (CH), Friesian (FR), Hereford (HE), Holstein (HO), Limousin (LM) and Simmental (SI) breed fractions were retained leaving 822,763 records. Also animals where less than 75 per cent of their breed fraction was known were discarded. And carcass weights of less than 150kg and greater than 550kg were also omitted.

"And we found absolute genetic correlations between the performance test measures, excluding FE traits, and the three carcass traits ranged from 0.004 to 0.33," said Mr Crowley.

Although standard errors were large, ranging from 0.10 to 0.16, they were considerably smaller than other studies with similar objectives. And most of the genetic correlations with performance traits estimated in this study were not more than two standard errors from zero with the exception of the correlation between CF and BW, so selection for increased BW is expected to have a negative effect on CF.

"With regard to the feed efficiency traits, CC and CW were both negatively correlated with FCR and this suggests that selection for improved, or lower, FCR will improve both CC and CW," explained Mr Crowley.

"Furthermore, CF and RFI were correlated indicating that selection for improved, or lower, RFI will yield leaner carcasses."

He also stressed that the standard errors of the genetic correlations were large similar to previous studies, attributable mainly to the relatively small datasets for the FE variables.

"Therefore, there may be merit in pooling data from international sources to increase the dataset size and potentially generate more precise estimates of genetic parameters," he added.

Calving Ease Proofs Rate Sires For Calving Performance

Calving ease proofs will help to reduce calving difficulties and their cost, although the heritability is low.

Having calving ease proofs available will help dairy producers to identify bulls that are genetically good (or bad) for calving performance. And this may be of particular importance when choosing bulls to use – or avoid – on heifers to reduce the number of difficult calvings and cut the costs associated with losses in production, fertility and potentially cow/calf losses.

So said the SAC's Eileen Wall, when she presented the findings of her team's project, to develop routine national calving ease evaluations for UK dairy cattle, to delegates at the 2010 British Society of Animal Science's annual conference, held at Queen's University, Belfast.

The ease of calving influences the economics of a cow/calf enterprise through increased calf death loss, increased labour and vet costs, reduced subsequent reproductive performance of the cow, potential loss of the cow, and reduced milk production. It's estimated that a slightly difficult calving costs approximately £110 and a seriously difficult calving costs between £350 and £400. "So a national calving ease evaluation would be a huge step forward for the UK dairy industry," said Dr Wall, explaining the rationale behind her work.

The project took data from UK milk recording organisations, and included producer-recorded calving ease (CE) data, as well as data collected as part of the progeny test scheme. Genetic parameters for CE were estimated, considering a direct and indirect effect.

Fixed effects in the model include herd year, month of calving, lactation number, calf sex and interaction between lactation number and calf sex. Age and percentage Holstein were fitted as linear regressions. And a random effect of service sire was fitted to estimate direct CE predicted transmitting ability

(PTA) and random effect of maternal grandsire fitted for indirect CE PTA, with a covariance between the two effects also fitted.

"The data spanned 15 years, but the majority of the data falls after 1999," said Dr Wall. "Overall, 84 per cent of calvings were classed as 'easy calving'. For first calving cows a total of 74 per cent calvings were classed as 'easy' and 85 per cent of later calvings," she added.

The genetic analysis showed that the heritability for calving ease was low (0.066 and 0.040 for direct and indirect effects respectively with a genetic correlation of-0.685). And there was no evidence of a genetic trend in either direct or maternal CE PTAs.

The across country genetic correlations for the multiple-trait across country evaluations (MACE) run for direct and maternal CE were in line with other countries and suggest that UK CE PTAs would be suitable for an international MACE evaluation. The genetic correlations between countries for the direct calving trait averaged of 0.80 and ranged from 0.619 (with Hungary) to 0.944 (with Canada).

The genetic correlations between countries for the maternal calving trait was a little lower with an average of 0.69 and ranged from 0.561 (with Hungary) to 0.839 (with France). These correlations with other countries are good, particularly for such low heritability traits.

Eating Quality Obtained Through Breeding

Marker assisted selection can be used in the future to allow beef breeders to select for meat quality to meet consumer demands. The low heritabilities estimated for meat quality traits, such as juiciness and tenderness, and the obvious difficulty involved in measuring these phenotypes, limits the effectiveness of traditional quantitative breeding. And making progress with these traits using direct measurement is likely to be time-consuming and costly.

So says Dr. Gill, from Edinburgh Roslin Institute. "For these reasons, we believe that the next step in the improvement of such traits is to investigate whether associations exist between meat quality traits and genetic loci, possibly starting from candidate genes.

"If significant associations can be found, then marker-assisted selection could be implemented. And this process has the added advantage of being able to assign breeding values to live animals, so that post-slaughter scoring is not necessary."

Sensory traits are known to be important to the consumer and will influence their consumption of meat, specifically beef. But these traits are difficult to measure and often require the use of taste panels to assess the complex parameters involved in the eating experience. Such panels are potentially a large source of measurement error, which may reduce the effectiveness of breeding programmes based on the data they generate. So the aim of Dr. Gill's study was to assess the quality of such taste panel-derived sensory traits as well as calculating genetic parameters and residual correlations for these traits along with a further set of traditional carcass quality traits.

The study examined a sample of 443 Aberdeen Angus-cross animals collected from 14 breeder–finisher farms throughout Scotland. To assess the quality of the taste panel measurements, three consistency statistics were calculated: panel-member consistency, or the extent to which an individual panel member varied in their scoring for a given trait over the period of the experiment; repeatability, or the consistency with which an individual panel member was able to score a trait on repeated samples from the same animal; and reproducibility, or the extent to which taste panel members agreed with each other when scoring a trait.

"These consistency statistics were moderately high, particularly for panel-member consistency and reproducibility, with values ranging from 0.48 to 0.81 and 0.43 to 0.73 respectively," said Dr. Gill. "Estimated heritabilities were low for most of the sensory taste-panel-evaluated traits where the maximum value was 0.16 for overall liking. Residual correlations were high between many of the closely related sensory traits, although few significant correlations were found between the carcass quality data and meat quality traits."

Young Sires Benefit From Genomic Selection

Using genomic information in beef animals increases the benefits to selection, particularly for young bulls and selection parameters that have low heritabilities such as residual feed intake.

There is a potential benefit that can be achieved from including genomic information in selection programmes in beef cattle. And this benefit is highest when selecting younger sires compared to older sires.

Just one of the findings of Carol-Anne Duthie and her team at the Scottish Agricultural College. Their work outlined the importance of the training population size and the effective population size as these constrain the potential benefit that can be achieved.

"Genomic selection is expected to be of particular benefit for traits that have low heritability and are difficult to measure or are only available late in the animals life or are sex limited," she told delegates at the 2010 British Society of Animal Science's annual conference, held at Queen's University, Belfast.

"Therefore, genomic selection may facilitate the inclusion of further traits, such as residual feed intake, in breeding programmes which are important to the efficiency of beef cattle." Genomic selection (GS) utilises information about the association of large numbers of single nucleotide polymorphism (SNP) markers located throughout the genome with phenotypic information such as growth rate. This has become feasible due to the availability of large numbers of SNP markers and the development of the bovine SNP chip. "And the aim of our study was to investigate the benefits of applying genomic selection to the terminal sire index of beef cattle considering training population size and different breeding structures," explained Dr. Duthie.

Selection index theory was applied to investigate the response of the beef terminal sire index under conventional selection or GS given the structure of the UK beef industry. Breeding objectives of the terminal sire index incorporate carcass weight, carcass fat score, carcass conformation score, gestation length and calving difficulty.

Currently recorded traits include birth weight, weight at 200 and 400 days, muscle score, muscle and fat depth, gestation length and calving difficulty. Parameters were obtained from UK beef genetic evaluations and information for the breeding goal traits were obtained from Amer et al. (1998). Selection strategies were derived from the structure of the beef industry, which was calculated from UK beef genetic evaluations data.

The effect of different breeding structures was investigated. This included selection based on young sires, where no progeny information is available, and selection based on older sires. Genomic information was included in the selection index model and accuracies of genomic estimated breeding values (GEBVs) are predicted based on trait heritability, number of phenotyped animals in the training population, the number of quantitative trait loci (QTL) underlying the trait and the effective population size (Ne).

"We found that, under conventional selection, the economic response is similar between the two breeding structures, however the slightly higher response (5 per cent) when selecting older sires is due to higher accuracy of breeding values due to the availability of phenotypic records on more relatives," said Dr. Duthie.

"Including genomic information increases the economic response to selection in both breeding structures, however the magnitude of the response in comparison to no GS is higher when selecting younger sires (up to 33 per cent increase) in comparison to the industry average (up to 21 per cent increase) where selection is based on approximately 40 per cent young sires, with a generation interval of less than three, and 60 per cent of older sires."

She added that the size of the training population influences the economic response that can be achieved when including genomic information, where the highest response is achieved with a training population of 5,000. "However, this is constrained by the breeding programme structure and Ne. Larger training population sizes had more impact when younger sires were selected. Furthermore, the rate of economic response was higher for Ne of 100 than 500."

Presented to the British Society of Animal Science Annual Conference, April 12 to 14, 2010, Queen's University, Belfast. Full details: Duthie C, Sawalha R, Navajas E, Roehe R and Roughsedge T: "Economic response that can be achieved from including genomic information to the terminal sire index of beef cattle."

To Improve Animals, You Must Improve Performance

Most beef producers still prefer to make selection decisions based on what an animal looks like, rather than a calculation of its genetic merit. However, this tradition may be undermining future profitability, writes Wayne Upton, Beef CRC extension specialist.

Most producers buy bulls the same way they have done for generations. Generally speaking, they will buy the biggest, fattest animal at a sale, and use their experience as third or fourth generation cattlemen or women to visually assess the genetic worth of an animal by what it looks like.

While tradition plays its part in keeping the status quo, another factor is that many commercial breeders do not have a strong understanding of how to use Estimated Breeding Values (EBVs) to improve the profitability of their selection choices.

They are unsure what the difference between tools like EBVs, which have been around for some 30 years, and genomics (the newer DNA markers and panels) are. For many, genetics is 'gobbledegook'.

EBVs are calculated from information on the animal's performance and the performance of its relatives and progeny. This information is used to make a prediction of the genetic worth of an animal. The newer genomic DNA markers are developed from an analysis of the minute differences in the genomic sequences of different animals which may point to genes that impact on a desired trait. In the future it is planned that this DNA information will also be used to increase the accuracy of EBVs. Although commercial breeders would like to have data on how well bulls perform for certain traits, they don't demand this information.

An example of some of the outcomes of selecting for traits using EBVs is profiled in the Southern Regional Combinations project. This project involved three sets of Angus bulls: one group selected on EBVs for intramuscular fat (IMF), the second for EBVs for retail beef yield. The third group was selected for progress in both traits. These bulls were mated to a random sample of Hereford cows and the progeny measured for IMF and retail beef yield at slaughter.

The results demonstrated:

1. Clear responses to selection of sires on EBVs for specific traits.
2. Excellent prediction of effects on carcase traits in the progeny.
3. Producers can select to improve more than one trait at the same time (even if they are negatively correlated).
4. Responses were consistent across a wide range of environments.

Collecting measurements on individual animals and their progeny is often raised as a significant issue for the northern beef industry, where cattle run in extensive production systems and where producers have managed herds over generations without measuring the genetics of animals. This is not to say it can't be done. In these extensive breeder regions, a renewed focus on heifer management, breeder performance and bull selection based on inherent fertility is imperative. The beef industry is not taking as much advantage of genetic improvement as other livestock industries, although it s at a similar level to the wool industry and the lamb industry is making improvement.

Basic Genetics

Genetics is used in the cattle industry to get a desired trait, for example-carcase yield or feed efficiency. This trait is influenced by two things.

Firstly the genetic makeup of the animal, but equally importantly, by the environment in which it is reared.

It is safe to say that each animal receives half of its genes from its sire, and the other half from its dam. However the random combination of these genes, means there is a widespread difference in the offspring. Some offspring may be exceptional, some average and some below average.

There are *additive* genetics and *non-additive* genetics, Mr Bullock told the audience of cattle producers from across the US.

Additive: These are predictable effects that get passed on from generation to generation.

Non-additive: These are dominance and genetic interactions, and are influenced by how the genes match up.

Crossbreeding has occurred for centuries. By crossbreeding, producers can give the offspring an advantage over the average of the parent breeds, because of non-additive genes and the way they match up. This advantage is known as heterosis.

There are two types of traits, qualitative and quantitative. Qualitative traits are usually controlled by one pair of genes, and are not easily influenced by the environment, said Mr Bullock.

An example of a qualitative trait would be coat colour. The coat colour of the animal will be determined by the dominant/or recessive genes carried by its parents. For example, black (B) is dominant to red (b). Animals with genes BB would be a homozygous dominant. Animals with Bb would be known as a carrier, or a heterozygous.

Red coated animals would be homozygous recessive-bb.

If two carriers mated, there is a 25 per cent chance that the offspring would be homozygous recessive, and 25 per cent chance it would be homozygous dominant. There is also a 50 per cent chance it would be a carrier.

However, if a homozygous dominant was bred with a carrier, there is a 50 per cent possibility that the offspring will be dominant (BB), and 50 per cent chance it will be a carrier (Bb).

Quantitative traits are controlled by many gene pairs, and the environment will play a major role in the outcome, said Mr Bullock. Quantitative traits are measured by heritability, the proportion of the trait which is controlled by additive gene action.

The majority of production traits are quantitative traits, said Mr Bullock.

Because of the numerous possibilities with quantitative traits, due to environmental affects, they can be difficult to improve genetically. More and more research is being done to improve these traits, as they are often economically significant ones such as carcase yields, fertility, feed efficiency etc.

Heritability measures the proportion of a trait that is controlled by additive gene action. It is the resemblance between parents and offspring, for a particular trait, due to genetics.

If a trait has low heritability (e.g. reproduction) this indicates that there is a low involvement of additive genes and/or that the environment has a much larger influence on the trait. High heritability indicates that additive genes play a relatively large role in the trait change, and so these traits are less affected by the environment.

Whilst fertility does have low heritability, traits such as birth and weaning weight, carry high heritability, as does yield per cent of carcase weight. Mr Bullock says that the heritability of the traits does vary between breeds.

Selection

Selection is the process of introducing the most desirable genotypes that will leave offspring that are beneficial to the herd and enhance business productivity and profitability. After several generations of selection for certain traits the more desirable genes become more frequent and the less desirable genes become rarer with the result that the overall genetic and phenotype merit of the herd increases.

Selection method indexes, such as estimated breeding values (EBVs/ EPDs) can result in rapid genetic improvement. Traits can be prioritised depending on economic performance, and single traits can be compared. The animals selected for breeding are those with the highest scores based on prioritised traits.

Whilst this is a reliable methods, visual assessment is still required, as not all traits can be measured.

Over time and before selection indexes, selection methods have definitely made a difference, said Mr Bullock. Looking across the breeds, there have been drastic changes in the genetic merit of most traits. This, Mr Bullock says, indicates that simultaneous selection for antagonistic traits is possible.

Measuring Genetic Merits

Measuring genetic merits is a complex task. There are number of different methods to help selection, including visual, estimated progeny differences, genomics and selection indices.

A trait is measured by taking the genetic influences and adding the environmental factors. However there is always an unexplained variation, said Mr Bullock. This variation cannot be measured. For example, if breeding with a bull that was born at a weight 75 lbs, his progeny's birth weight could vary between 55 lbs and 95 lbs.

The more genetic data available, the less the unexplained variation will be.

Concluding, Mr Bullock said that selection should be made on traits that economically impact individual herds.

What is An EBV and How can it Help You?

Estimated breeding values (EBVs) are an essential selection tool, which are underused in the UK, says William Haires, Nuffield scholar and Pedigree/ Commercial beef breeder.

What is an EBV?

An EBV is a value which expresses the difference (+ or-) between an individual animal and the herd or breed benchmark to which the animal is being compared. EBVs are reported in terms of actual product e.g. days, kg of weight or mm of fat depth, etc.

There are two main performance recording services operating in the UK.

- The Australian Breedplan programme is used by many beef breeds in the UK and operates through Pedigree Cattle Services.
- Signet/BASCO is operated by Edinburgh Genetic Evaluation Services (EGENES).

Both systems produce EBVs for many of the same traits, with their databases being accessed via the respective breed societies. Sale catalogues that have EBVs included for each animal should also have an explanation of the traits and a chart showing percentile bands, depicting where the individual animal ranks within the breed depending on its performance for the traits in question.

Contrary to common belief EBVs are not a massive waste of time and money but an estimation of how the progeny of an individual should perform! The purpose of EBVs are to bring all the cattle within a herd or breed onto a common platform so that those making selection decisions can know they are comparing like with like regardless of the system or location they are produced in. It is important at this point to remember that each parent contributes half the resulting calf's genetic make-up.

This performance estimation measures the parts of the animal that the eye. An animal's physical appearance (phenotype) is determined by two components, its genetics (genotype) and non genetic or "Environmental" influences. EBVs measure this genetic component and allow cattle within their respective breeds to be compared, excluding their management and Environment. The key point here is what you see is not necessarily what you get!

The environmental (non-genetic) influences include grass quality, disease burden, parasites, supplementary feeding, management ability etc. The environment component is a part that herd keepers have influence over and the ability to compensate for or enhance. Despite the environmental influences, the genotype remains the same and parents do not pass on to their progeny the environmental effects that have influenced them.

EBVs draw information from a number of sources in addition to the animal's own performance. The basis is comparing individuals within a contemporary group, one to another, minimising non-genetic influences. The system also compares contemporaries in other herds where genetic linkages exist through siblings and ancestors. This is a very simple description of what is involved to outline the principle. The actual calculations are much more complex and beyond anything that a farmer needs to know or understand. Linked traits are those which share common connections with each other. An example of this is that where 600 day weight increases so will birth weight.

Heritability is the term used to describe how strongly a characteristic is transferred from parents to progeny and is rarely considered. It is the proportion of an animal's production that comes from its genetics and varies between traits. The higher the value the faster genetic improvement can be made and positive benefits observed. In general terms, maternal traits are poorly inherited and growth traits are moderate to highly heritable. The old adage of cattle performance being "80 per cent feeding and 20 per cent breeding" is not strictly true, but the breeding component will determine how the non genetic component is used. At low to average levels of performance, feeding and health will have more of a contribution to an animals overall performance but for those trying to maximise returns the right genetics are vital. A typical example could be milking ability, where if the genetics for milk are poor, feeding will not make a significant difference, the cow will just get fatter.

Accuracy

No two animals are the same, so the more information that can be collected, the more one can account for the variation between individuals, and so the overall risk of using them can be reduced.

EBV data must generally reach a certain level of accuracy before it can be reported to ensure the information is as relevant as possible and truly reflects the potential performance of the animal in question. Some traits such as growth will not be reported until they reach an accuracy level of 40 per cent, which is typically at the lower range of reported accuracies (these can be up to 99 per cent). Traits which are more difficult to measure such as carcass and fertility related can be reported at much lower levels due to the limited information available, requiring the user to make a judgement decision.

PRODUCING GOOD QUALITY EBVS

Throughout a performance recorded animal's life, measurements are taken which are used to produce its EBVs. These start with a birth weight and calving ease score. These measurements are supplemented by weights at 200 days, 400 days and 600 days as well as scrotal circumference (bulls only) and ultrasonic measurement of back-fat, rump fat and muscle depth. At weaning time, dam weights will also be collected to contribute to the mature weight EBV.

In the future, genomic data may also be incorporated for the difficult-to-measure traits such as carcass and reproductive characteristics. Disposition data is collected by some overseas breeds and a docility EBV produced to measure their temperament based on "flight time". In a study carried out in Australia it was found that "flighty" cattle were gaining 0.4kg/day less than their docile contemporaries, a substantial amount in today's challenging climate. Signet will be launching a docility EBV in 2011, leading the way in the UK.

In addition to the information collected from an individual, the data from siblings and other family members make a contribution. There are many factors involved in the production of quality EBVs but it is useful to mention a few of the factors that lead to poor quality EBVs as well. The most important principle of performance recording is that of grouping the cattle together for the comparison. This is the biggest challenge for the system to overcome and depends on accurate record keeping and submission of raw data.

Where the same sire is used on the same cows year after year or where there is only a small number in the management group there is evidently a lack of genetic diversity. This limits the amount of variation that can be accounted for.

Management for Better Quality EBVs

To produce good quality EBVs it is important that the herd has a compact calving pattern, with the cows run in reasonably large groups and that any data from that group is collected on the same day. Within this group at least two different sires should be used each year to allow the system to compare the performance levels between these sires' calves.

Ideally the bulls should be a mix of a proven sire and a young or unproven sire from the recorded animal itself. This information depicts the type of genetics that an individual actually possess because it helps estimate the genetics that it will pass to its progeny.

Any animal that may have been sick, fed for a show or sale, used as a sire or introduced to a new group should be identified to ensure it is not compared on a "like for like" basis with the rest of its group. In small herds the use of AI bulls will lead to rapid improvements in their EBVs and accuracy.

This applies to all sizes of herds. In small herds the risk may be to use a bull retained from the herd. His use is limited to only a small number of females each year meaning the chance of getting any meaningful information is negligible. Small herd owners need to evaluate the cost of keeping a bull compared to AI, given modern heat synchronisation techniques which make the process much simpler and more accurate.

The main principle must be to record all calves, the good the bad and the ugly. Selectively recording animals will deny a true picture of what is happening in the herd.

It will give a distorted base set unreasonably high or it may undervalue better sires whilst overvaluing lesser sires. Accuracy in measuring weights and correctly recording dates of birth can make a substantial difference.

Estimating weights is not accurate enough and whilst not having a significant effect on the overall group (as the estimate will likely be variable), 50kgs of error either way can have a much more serious impact on the individual, adversely affecting related traits especially those with negative relationships. Accurate honest weights are vital to many processes on farm including medical treatments and pre sale checks. A weigh scale is the most important piece of equipment on a livestock farm.

Popular Misconceptions

It is often asked why a good animal does not have good EBVs, or vice versa:

- Firstly an EBV is a performance estimate or prediction for the progeny of a bull or cow and not of their own performance.
- Secondly the animal will most likely have had its environment influenced in some way. It may have been taken away from the rest of its group, been suckling another or several other cows, had creep feed, been housed earlier or weaned later or some other such practice which has compensated for its lack of genetic potential.
- Below are two of the reasons why some cattle look outstanding but their EBVs do not reflect this. The questions which should be asked at this stage are "What would this animal have looked like without additional management?" and "What will his calves look like without similar expensive management?"
- Where a calf with superior genetics suffers a setback such as disease affecting it or its mother, the opposite effect can occur. This animal still has all the potential and genetics to produce great progeny but may not look as good as it could or as good as its contemporaries. This highlights the danger of selecting by eye alone, as this animal could be inadvertently rejected.
- Animals that have been imported from overseas can suffer from poor, low accuracy EBVs. With limited or no information about its past performance, its EBVs will unfortunately reflect this until sufficient data is collected and processed, in its new home country.

Concern is often expressed about the possibility of a person exaggerating the weight of an animal to increase its EBV. This could happen but it would be unwise.

There are checks and balances within the EBV system which queries animals that are too far ahead of the group and secondly, as progeny are analysed its failure to perform to expectations will be noticeable. Animals that are outside the boundaries are flagged as "outliers" for further investigation.

Anyone under the illusion that he can beat the system and fool the process is wrong. He may get away with it for one or two calf crops but the cattle will fail to live up to their expectations with their EBVs deteriorating rapidly when they go into production. The biggest victim in this fraud is the person recording the data as he is fooling himself and destroying his reputation. Breed societies need to act aggressively in any cases detected to protect their members and the integrity of the breed.

Selection Indexes

Selection indexes take the hard work out of knowing how much emphasis should be placed on each of the available EBVs when making breeding decisions. An index gives a single EBV that reflects the value of an animal in financial terms. Indexes allow balanced selection as they apportion the amount of selection pressure that needs to be applied for growth, maternal, carcass and fertility traits to give the most profitable herd over the long term. Typical production parameters, prices and production costs underlie each index. The next part involves steps that combine economics with genetics. Financial values for performance measures are calculated for each breed's production and market. Using genetic theory these financial values are used to calculate appropriate weightings for the EBVs currently available.

Indices exist for different markets such as terminal index and self replacing index and provide a convenient way of speeding up a search. Choosing the highest index animal will not necessarily result in a suite of traits that will suit the herd and compromise will likely be needed so it is still essential to look at the EBV for each trait and determine if they are appropriate to ultimately satisfy the previously identified needs.

How much is Genetic Improvement Worth?

At this point it is worth using an example to show how value can be calculated using EBV data. As already mentioned Indexes are a financial value. Using the example of Dorepoll 1 10H Headliner (appendix 2) who has a Self Replacing Index (SRI) of £36.00 compared to the breed average of £24.00, there is a difference of £12.00. In a herd producing 100 calves over his breeding lifetime this equates to £1200. Compare this bull to a bull with a self replacing index of £10 and the extra value to the business immediately leaps to £2600!

Irish Cattle Breeders Federation

The Irish Cattle Breeders Federation (ICBF) has developed a product named "Herdplus". This system produces management reports for both the pedigree and commercial herd owners. EBVs and indexes are also produced for both types of herd meaning that even commercial suckler cow herds have performance recording information available for all their cattle. The system draws on information from markets, abattoirs and data collected by the farmer. Hopefully the UK will be able to develop a similar system in the years to come.

Improving Performance In Suckler Herds Through Breeding Plans

Breeding is the key to a successful future for suckler herds, said William Haires, Nuffield scholar at the British Cattle Breeders Conference. Charlotte

Johnston, TheCattleSite Editor reports. Embarking on a Nuffield scholarship in 2010, Mr Haires visited the USA, New Zealand, and Australia to find out what he had to do in order to run a successful suckler herd into the future. He specifically looked at selection techniques and performance improvements, to establish the practical benefits and application of estimated breeding values (EBVs) and Gene Marker technology.

The UK has been known as the genetics capital of the world, introducing Artificial Insemination (AI). "The industry once dared to be different," he said. "Now the UK has been left behind by nations who used to look to up to us.

"Suckler herd management must change," he stressed.

With costs of production increasing, new practices and existing ones need to be evaluated. There is no opportunity to 'get it wrong' now, especially as the value of subsidies has fallen significantly.

Judging cattle performance by eye alone, is no longer a way to run a profitable business. Improving the genetic capabilities of a herd is the cheapest and most beneficial investment a herd owner can make, Mr Haires said.

He criticised the industry for moving towards continental and other breeds, which he says have let breeders focus on correcting faults in the breeds, rather than look at improving efficiency.

"Farmers have been trying to make their farm fit the system, instead of choosing a system to fit their farm," he said.

"Because breeders have become so busy treating the symptoms, we have forgotten what the problems ever were."

UK cattle have so much undervalued potential, said Mr Haires, and producers just aren't using the resources available. With rising feed costs, why aren't producers establishing breeding programmes that allow for greater use of grass-our biggest asset, he asked.

"Ultimately we want to breed cattle that perform well, are efficient and improve profitability."

So What Makes a Good Breeding Programme?

Through travelling abroad, Mr Haires has seen the value that genetic improvement has on breeding cattle to work more effectively with the resources already available.

The following are what Mr Haires considers vital when creating a breeding programme:

Market requirements: Are cattle being produced for meat or breeding? Live or dead market? Opportunities exist to produce cattle for existing or new premium markets including those based on breed, provenance, eating quality or commercial replacements.

In order to understand market requirements, knowledge is key. "It is vital that better relationships within the supply chain are developed to share requirements and knowledge which benefit everyone involved," he said.

"The current lack of shared information is restricting producers' ability to supply what the retailer and consumer demand because they simply do not know what is required."

Resources available: What land is available, what systems suit this land, environmental practices in place, buildings available.

This list is not definitive-as each farm is different and every farmer needs to evaluate his own, but whatever the resources are, they will determine the types of stock that can be kept, the potential markets that can be reached and the extent to which improvements can be made.

Herd records: This is to aid identification of cattle within a herd, and will help measure and identify individual traits at a later date.

Personal and Financial Expectations

Within herd demands: Is the herd closed, what vaccinations are required, replacement rates, stocking rates, herd docility, intervention between humans/ cattle etc.

Selection tools: This is important to assist with breeding herd decisions and culling options. Selection tools could be weighing scales, EBVs or other genetic markers. Mr Haires says that cattle performance must be measured.

"The old adage of cattle performance being 80 per cent feeding and 20 per cent breeding is not true," he said. The breeding component will determine how the non genetic component is used, when trying to maximise returns, the right genetics are vital.

EBVs

EBVs should be the minimum criteria used for selecting breeding bulls, semen for AI and pedigree females for breeding, Mr Haires said.

Mr Haires found that the most successful herds he visited abroad were ones that had completely embraced EBVs. He said that whilst getting the EBV for the bull right, the bull is only half the story. Therefore cows must also be selected on EBVs.

He found that the US were struggling to get used to EBVs (called EPDs in the US), but that progressive farmers were using them successfully.

An estimated breeding value, is exactly that, he said. It is an estimation or prediction for the animals progeny, not its own performance.

Mr Haires suggests that EBV targets should be set with the system that suits the farm in mind. Bulls/cows should then be looked at and it is vital that they fit the criteria set. This way a package of traits is acquired.

Using animals with higher indexes than the average for that herd will have significant benefits for the herd, increasing the value of the business.

DNA Based Genetics

Incredible process is being made with regard to genomics. Whilst the bovine genome has been mapped out, there is a vast amount of research still needed to identify what each single nucleotide polymorphism (SNP) (found in the DNA) controls and influences.

The use of genetic markers, allows desirable traits to be recognised in cattle at one day old. Their greatest use comes with traits that have a low heritability, are difficult to measure, cannot be measured until the animal has produced progeny (by which time it may be too late), where the animal may have to be slaughtered or where the trait is not routinely measured.

"If the trait can be measured using conventional performance recording there is no need to use anything else. If there is no other method of measuring a trait, cost is not prohibitive and it is a benefit to the business, it may be worth considering genomic analysis as part of a comprehensively designed plan," said Mr Haires. Mr Haires farms 220 acres near Belfast, Northern Ireland. The farm runs pedigree Hereford and Angus herds, and a commercial suckler herd based on Ayrshire/Friesian X Hereford cows mated to native breed sires (Aberdeen Angus).

Mr Haires is currently looking at shortening the calving time, which is currently eight weeks. By shortening this, he hopes to be able to turn out animals earlier, at the start of May.

In the future, he is also looking at introducing a whole farm approach, including cell grazing. This, he says, will tighten control of the cows grazing, so they would eat areas that would usually be neglected.

One intensive grassland system that he saw in New Zealand was achieving a daily liveweight gain of 2kg per animal.

Concluding Mr Haires said: "The UK has some of the best conditions in the world to produce beef, with the use of EBV's and genomics, the future for suckler beef production is an exciting and profitable one."

Beef Breed has a Considerable Effect on End Value

Breed and sex differences in carcase distribution can have a considerable effect on the overall value of beef animals due to differential pricing of various cuts, according to studies carried out in Edinburgh and Bristol.

Steers have a greater proportion of forequarter (shin and clod) and lower proportion of hindquarter (topside, rump and sirloin) in their carcases compared with heifers.

Just one of the findings of a study, carried out by scientists in Bristol and Edinburgh, to quantify the proportions of various primal cuts within beef carcases from both Aberdeen Angus cross (AAx) and Limousin cross (LIMx) steers and heifers slaughtered through a commercial abattoir.

"This may be a function of circulating hormone levels," the Scottish Agricultural College's Jimmy Hyslop told delegates at this year's British Society of Animal Science's annual conference, held at Queen's University, Belfast.

Carcass balance, or the proportion of overall beef carcase weight that is present in various primal cuts in both the hindquarter (HQ) and forequarter (FQ) segments, has a considerable effect on the commercial value of each carcass.

So Dr. Hyslop's team took nine AAx steers and seven each of AAx heifers, LIMx steers and LIMx heifers from a range of dam types. One side of the slaughtered carcases were cut into a total of eleven commercial primals, vacuum packed and frozen at-20°C.

Although the 11 primals were further sub-divided for other experimental procedures, the weight (kg) of each commercial primal was then expressed as a proportion of the total matured carcase side weight (g/kg carcase weight) for the purpose of the 'carcase balance' study.

"And we found that LIMx animals had a higher proportion of hindquarter compared with AAx animals with the differences being larger between LIMx and AAx steers compared with the difference between LIMx and AAx heifers," said Dr. Hyslop.

"Proportional differences among 'carcase balance' can have a considerable effect on the overall value of animal breed types and sexes due to differential pricing of the various primals," he added.

Production and/or Profit?

Speaking at the recent Beef Improvement Federation Research Symposium and Annual Meeting, held in Columbia, Matt Spangler, from the University of Nebraska-Lincoln looks at focusing breeding objectives by selecting for profitable genetics, not high production genetics.

Steep increasing genetic trends for growth traits (weaning and yearling) and mature cow weight can be seen in many breeds but perhaps more alarming are those producers that have dramatically increased the genetic potential for milk production in their cow herds.

Conditional on the assumption that the Beef Cattle Industry is a For Profit organisation, then it would seem logical that profit (Revenue – Expenses) should drive our selection decisions. In order to actually do this, knowledge of environmental constraints, genetic antagonisms, and the selection tools that have the potential to measure profit are critical.

Environmental Constraints

The development of an obtainable breeding objective begins by clearly identifying environmental constraints and marketing goals.

If feed resources are limited in a stressful environment then selection for increased extreme output (high growth, milk, and red meat yield) could have negative impacts on the ability of cows to be successful breeders without the need for large quantities of harvested forage. The beginning of a profitable breeding objective is identifying what the environment will allow you to produce, at least until we have tools to apply direction selection to traits of adaptation.

Crossbreeding

Commercial producers who have not yet adopted crossbreeding are a burden to the beef industry.

We know that the two primary benefits of crossbreeding are complementing the strengths of two or more breeds and heterosis, neither of which create trait maximums. If we think about it simplistically, crossbreeding for a trait like weaning weight leaves us with a calf crop that is better than the average of the parental lines, not better than both parental lines.

Crossbreeding, if done correctly, seeks to optimise many traits through complementing breed strengths and produce animals that are better than the average of the parental lines that created them. The best tool that the commercial cattleman ever had is based on optimisation, not the production of extremes. So, it would stand to reason that within breed selection should have the same goal, optimums and not maximums.

Genetic Correlations

Unfortunately, all traits that might be included in a breeding objective are not independent of each other. Sometimes this is beneficial as we see a favourable correlated response, and other times these genetic correlations pit revenue against cost. A good example of this comes from the suite of weight traits. Depending on the targeted marketing endpoint either weaning weight (WW), yearling weight (YW) or carcase weight (CW) become a source of revenue and all are related to a major factor influencing the cost of production, mature cow weight (MW).

Although it is not intuitive, literature results show that of the immature traits, WW has the highest genetic correlation with mature cow weight.

Care should be given not to focus solely on the revenue portion, sale weight, but rather optimising input costs associated with mature weight and revenue sources from calf sale weight. The mature sale weight, CW, shows a strong and positive relationship with MW and again care should be taken to optimise selection between the two.

Selection for Decreased Input

In order to mitigate genetic antagonisms in an effort to select for profit, economic index values become the tool of choice. A bio-economic index (H) is simply a collection of EPDs that are relevant to a particular breeding objective, whereby each EPD is multiplied by an associated economic weight.

A high index value does not necessarily mean that an animal excels in all EPD categories given that superiority in trait can compensate for inferiority in other traits depending on how the EPDs are weighted in the index. A high index value should be thought of as excelling in the ability to meet a breeding objective.

The majority of economic index values are rigid (i.e. not catered to individual enterprises). A much more desirable method would use individualised index values where the bull with the highest index value may differ from one herd to the next, depending on how the animal fits the specific needs of each enterprise. While this would lead to more accurate identification of parents for the next generation, it becomes a challenging metric to use for advertisement purposes in the seedstock industry, which probably explains why this more desirable method of multiple-trait selection has not been exploited by the majority of breed associations.

New and Improved Tools

Genomic tools hold the potential to provide predictions for hard to measure traits that focus on input costs such as feed intake. Ideally, genomic predictions for feed intake would be incorporated into an economic index as a key component of input cost. However, accurate genomic predictions will require phenotypes.

The improvement of existing phenotypic databases for traits is also needed. It is critical that seedstock producers routinely turn in mature cow weights along with body condition scores to further aid in selecting for optimal weights and the development of tools such as the American Angus Associations Cow Energy value index ($EN) and the Red Angus Associations Maintenance Energy (ME) EPD. This will require participation in Whole (or Total) Herd reporting, a very necessary process for complete data collection and the development and delivery of genetic prediction tools. Trends are rarely flat, as an industry we have measured ourselves by steep lines in one direction or the other. From a seedstock perspective this may have been perceived as necessary in order to differentiate themselves (either as breeders or as breeds) from others in the market place.

Clearly identifying your production environment and realistic production goals given that environment are critical. Profit lies in the optimisation of expense and revenue and optimisation is always more challenging than maximising outputs or minimising inputs. It will require more effort, detailed financial records, and a structured breeding objective that builds a cow herd based on optimum values and not extremes. One final thought, extremely low maintenance cows will push the lower threshold of what is biologically possible for weight and produce virtually no milk. High output cows will represent the other extreme, weigh more than most mature bulls and milk heavier than the best Holstein. Both excel in some measure of the profit equation (i.e. lowest cost or highest revenue) but neither promises to be profitable.

Genomics to Improve Your Business

Today's beef cattle are the result of selection processes that have been refined for centuries. Producers have developed a strong genetic base and are designing cattle to meet the production criteria needed to fit their individual environments, says North Dakota State University beef specialist, Kris Ringwall.

Producers simply do not round up rogue cows and calves, select a calf for harvest and then invite the neighbors over for supper. Selection processes allow producers to get a feel for the genes a bull carries without having to gamble on the bull's outward appearance.

Two selection methods are readily applied today. The first is the use of expected progeny differences (EPDs). The second is the development of the newer technologies that involve analysing DNA, commonly referred to as genomics.

EPDs help the beef industry evaluate the genetics of a bull. The outward appearance, commonly referred to as phenotype, of a bull is a product of the bull's genes and how those genes interact with the local environment. Then it certainly is reasonable to make every attempt to get the best estimate of what genes are present in each bull.

For years, the environment tended to get in the way. The nice, shiny, perhaps overconditioned bull did not always sire shiny, well-grown calves in the fall.

Producers experienced a major let down and realised money was wasted on an overinflated bull. Published EPD values for each bull are intended to avoid buying mistakes.

EPDs inform producers of the expected performance of the bull's offspring relative to other bulls in the breed. Producers literally buy this expected performance when purchasing bulls with EPD values needed for their respective breeding programmes.

The EPD concept and the ability to improve a herd are based on an understanding of accuracy expressed as a value from 0 to 1. The accuracy is an indication of the amount of data going into the EPD estimate.

Accuracy includes the history of an individual's own performance data. Accuracy also includes performance data available within the pedigree of the bull and then progeny performance of the bull's offspring.

Now the good news or better news, depending how one wants to express it. Newer EPD estimates now have the capacity to add genomic information based on the analysis of the actual DNA of a bull.

The industry has been buzzing about these new analytic procedures, but a lot of chitchat would imply that the clarity of what a DNA analysis might add to a producer's beef cattle breeding program is unclear.

Producers generally grasp the concept that if a particular gene has been identified as bad and the makeup of the gene is known at the molecular or DNA level, then samples can be submitted and evaluated to see if the gene is present.

If it is, the animal can be culled. If not, the bull would not have the trait to pass on.

The same is true for published EPD traits. As mentioned earlier, EPDs are estimates that utilize the data of a bull's DNA.

This DNA, combined with management and environmental effects, is responsible for the phenotypic appearance of a bull.

The current excitement stems from newer genomic techniques that more directly link the phenotype of a bull to those long strands of DNA that stay very well hidden within chromosomes. These long strands of DNA ultimately are passed on at conception to create new calves. No, we are not necessarily creating more traits. Through genomics, we are adding to the knowledge we gain from the current EPDs available on the bulls we buy.

And that leads to increased accuracy in our selection. And that leads to better placed money when we buy bulls. And that leads to better performing cattle within the environments we are raising them in.

And that leads to more opportunity to make money. And that leads to a better beef business. And that leads to making us all happy!

Maximise Value of 2011 Calf Crop

Producers soon will be weaning calves and finalising marketing decisions. North Dakota State University Extension Service beef cattle specialist, Carl Dahlen, discusses how producers can capture the most value from their calf crop.

"For producers to capture the most value from their calf crop, they need to follow a carefully planned marketing strategy," says Carl Dahlen.

Here are four strategies to maximise calf crop value this year:

- Start with a uniform group of calves.
- Implement a sound herd health program.
- Consider age and source verification.
- Talk with your auction market representatives.

The sale price almost always is greater for a large group of high-quality, uniform calves than similar-quality calves brought to the livestock auction in smaller groups or similar calves with more weight range variation.

High-quality genetics and groups that are uniform in colour, frame, muscling, flesh and weight range are essential to add value to a group of cattle. To improve calf uniformity, concentrate on purchasing high-quality breeding stock, managing breeding seasons to create a tight calving window (consider a 45-day breeding season) and possibly culling late-calving cows.

"Implementing a sound herd health programme is the easiest way to add value to the calf crop, but it has to be done correctly," Mr Dahlen advises.

"Calves must receive the correct vaccines at the correct time."

Vaccines calves receive as they walk onto the truck do not have a large impact on the price buyers are willing to pay because most cattle will be revaccinated upon arrival at a feedlot. However, producers who vaccinated calves two to three weeks prior to weaning and followed all Beef Quality Assurance (BQA) procedures are more likely to be rewarded with higher prices.

The type of vaccine cattle receive also can impact the price. Cattle feeders are looking for calves that have received a modified live vaccine. Many feeders are not happy with the protection from a killed vaccine.

Recent sales have shown price swings of $5 to $8 per hundredweight for the same quality of calves, depending on vaccination status, with higher prices paid for vaccinated calves. Consult with your veterinarian about the vaccination strategy appropriate for your herd.

Even one year ago, auction market staff had not received a lot of requests for age and source verified calves or special age and source verified sales.

"However, 2011 is a different story," Mr Dahlen says.

"This year may prove to be the most beneficial year to date for producers who age and source verify their calves. As long as Japan restricts imports to less than 21 months of age, the demand for age and source verified calves will remain strong. But the process for verifying calves does not happen overnight."

Producers should be proactive, initiating and following through with the age and source verification process. They need to have records in place and start the paperwork well in advance of the sale to properly verify the age and source of calves. Mr Dahlen recommends producers talk to a third-party certification agency to learn if their calves and records qualify for age and source verification. Similar organizations can help producers certify that their calves are suitable for natural and organic markets.

"I encourage producers to spend time evaluating production costs and returns to ensure that they are obtaining a sufficient premium when marketing under a natural or organic label, compared with managing and marketing calves in traditional markets," Mr Dahlen says.

Auction market representatives can be great resources for producers with questions about improving the value of their calf crop. The market representatives deal with questions about feed yard preference for calves, status of calf supply and demand, whether feedlots are running at capacity and the latest market trends daily, whereas many cow-calf producers evaluate these questions once a year.

In addition, auction markets may offer their customers alternative marketing avenues, such as special sales, Internet sales and video auctions.

Producers should remain in contact with auction market staff and be flexible in their marketing strategy, according to mr Dahlen. Consign cattle well in advance of the sale and let the auction market know as much about the calves as possible (for example, what colour they are, when they were born, when they were vaccinated and whether they are age and source verified). The more information the auction market has, the better job it can do of marketing those calves.

In addition, producers should ask auction market representatives about other ways to improve their cow herd and calf marketability.

"No silver bullet will make you consistently top the sale each time you sell calves," Mr Dahlen says.

"Concentrate on creating a uniform set of calves with cows that work in your environment. Manage your herd with good BQA practices, vaccinate in a timely fashion, age and source verify if the market is rewarding the effort, and remain flexible with marketing strategy and dates."

Make Sure You Know What Your Gross Margins Are

When a producer understands gross margins, he or she can start to appreciate what dollars are available to work with, says Kris Ringwall, North Dakota Beef Extension.

A point of clarification is in order. Assessing income in a cow-calf enterprise is more than just the price received for the calves. The impact of making change also needs to be measured. If the impact of the change is measured only in calf value or dollars received per pound of calf sold, the measured impact is not the most accurate.

The assessment of the health of a beef enterprise is critical. If the current managerial efforts are generating the desired outcome, pen and paper should be able to track the effort. Did net return go up and did the desired managerial effort have a positive outcome? The measuring stick that is most beneficial is the calculation of gross margin per cow. That value is not the same as the calf value per cow. That is because there are multiple sources of income within a cow-calf enterprise.

The natural and appropriate tendency is to focus on cost as one evaluates a cow enterprise. However, if a major change in management is made such that the historical income opportunities also have been modified, then both the cost and income side of the operation must be assessed properly to understand fully the results that are positive, neutral or negative.

That is why the call for a point of clarification. Gross margins per cow should be discussed as a finite value. To change the value (increasing or decreasing gross margins), the dollar change is not directly connected to a particular dollar value in terms of price per pound of calf or any other livestock sale. It is a dollar value projected by the composite sale of all cattle products. This is the overriding business principle in using gross margins.

When a producer understands gross margins, he or she can start to appreciate what dollars are available to work with. How those dollars are obtained depends on what livestock are available to sell.

According to Jerry Tuhy, Bismarck State College farm business management instructor at the Dickinson Research Extension Centre, gross margin accounts for the purchase and sale of all calves, cull cows and bulls, plus the animals transferred in and any overall changes in cattle inventory.

The North Dakota Farm Management education programme, along with FINBIN data from the Centre for Farm Financial Management at the University of Minnesota, indicate that, since the turn of the century, per cow gross margins have averaged $501.42.

To get to that number, questions based on dollars per cow in the cowherd need to be answered. The answers will reflect the FINBIN estimates for the years 2000 through 2010. What was the value of calves sold per cow ($180.96)? What was the per-cow value of the calves transferred out of the cow-calf enterprise to a backgrounding or feedlot operation ($358.85)? What was the per-cow value of cull sales that includes bulls, cows or young breeding stock ($99.30)? What other miscellaneous sales of beef were accounted for per cow ($5.74)?

As a producer, if I stop there and add up the income per cow, the total sales income per cow would be $644.85. Keep in mind that, even though this is the average value of sales per cow, the average value per pound of calf sold cannot be calculated by this number. In fact, even if one looks at only the value of calves sold and calves transferred out, the average value per pound cannot be determined by these numbers.

It is critical to think gross margin concepts instead of value per pound of calf. However, we still do not have the gross margin calculated because the cowherd needs to be replaced. We have old cows going out and new cows coming in. Therefore, what was the cost per cow of all purchased breeding stock ($77.75)? What was the cost per cow of breeding stock transferred into the breeding herd ($69.07)? Finally, did the inventory within the herd change at all? If so, what was the value of that change ($3.38)?

Right now, all the numbers for calculating gross margin are available. If the sale numbers are added up, plus inventory change, one gets a gross margin of $648.23. To offset the cost of replacing the herd, one needs to subtract the cost of purchased or transferred breeding stock. If one does that, the average gross margin per cow since the year 2000 is $501.42.

THE BAYESIAN CONTROVERSY IN ANIMAL BREEDING

Frequentist and Bayesian approaches to scientific inference in animal breeding are discussed. Routine methods in animal breeding (selection index, BLUP, ML, REML) are presented under the hypotheses of both schools of inference, and their properties are examined in both cases. The Bayesian approach is discussed in cases in which prior information is available, prior information is available under certain hypotheses, prior information is vague, and there is no prior information. Bayesian prediction of genetic values and genetic parameters are presented. Finally, the frequentist and Bayesian approaches are compared from a theoretical and a practical point of view. Some problems for which Bayesian methods can be particularly useful are discussed. Both Bayesian and frequentist schools of inference are established, and now neither of them has operational difficulties, with the exception of some complex cases. There is software available to analyse a large variety of problems from either point of view. The choice of one school or the other should be related to whether there are solutions in one school that the other does not offer, to how easily the problems are solved, and to how comfortable scientists feel with the way they convey their results.

Animal Breeding & Genetics

The animal breeding and genetics section works to provide new scientific discoveries to age-old livestock production problems to help producers and consumers.

Currently several courses are available to undergraduate students interested in animal breeding and genetics, and both master's and doctorate degrees are offered in animal breeding.

The field of animal breeding and genetics research is more exciting than ever before, with projects such as bovine gene mapping and DNA sequencing. Using state of the art tools and facilities, the researchers at TAMU are able to contribute to the field of animal biotechnology on a worldwide level.

Animal Breeding and Genetics

Farm animal breeders have two core activities. By using selection techniques, they develop improved varieties of sheep, cattle, chicken and so on. And by accelerating reproduction they either produce, or assist the farmer to produce, large numbers of offspring of these varieties. In connection with these core activities, breeders undertake ancillary research on biotechnologies such as Artificial Insemination (AI) and collect data on animal productivity and health. For much of the twentieth century, the goal of breeding was improved production. Breeding targets included high milk yield in cows and large, fast-growing muscle in chickens farmed for meat. Recently, however, the emphasis has changed. Breeding programmes now aim to produce animals that are more disease resistant, enjoy higher levels of welfare, and provide good quality food.

Ruminants (Cattle, Sheep, Goats)

The cooperatives can provide their members with semen from a wide range of breeds, such as Holstein-Friesian, Jersey, Montbéliarde or Nordic Red breeds. Breeding organisations pursue research on AI and embryotransplantation, and undertake data registration (e.g. milk-recording and family information) for their members. Breeding information on dairy bulls is collected from around the world by the International Committee for Animal Recording (ICAR) so that international comparisons can be made.

AI use in ruminants varies a great deal. In dairy cattle in the Nordic and West European countries it accounts for nearly 100% of inseminations. But in sheep and goats, and in some Southern European countries, it is rarely used.

Countries in which AI is the main method of dairy cattle insemination use between one million and seven million doses, or 'straws', of semen each per year. (Precise figures depend on the scale of dairy farming in each country.) Between 1 and 3 artificial inseminations are normally needed per conception, and pregnant cows generally carry a single calf.

AI in beef cattle varies depending on the country and the breed. Overall, however, it is used less in beef cattle than it is in dairy cattle. Where conception is achieved through natural insemination, one beef bull normally serves around 25-30 cows. In sheep and goats insemination is usually natural. In sheep, one ram serves at most 30 ewes. Embryotransplantation is used commercially in cattle. Globally, just over half a million cattle embryos were transferred in the year 2000. In total 20% of these were European and 42% were North American.

The main European users of embryotransplantation were France, Germany, the Netherlands, the UK and Italy. In 2000 a total of 440 sheep embryos were transferred in vivo in Europe. In the same year 145 goat embryos were transfered.

Poultry (Chickens, Turkeys, Duck, Geese)

A small number of privately owned poultry breeding companies provide Europe and the rest of the world with breeding stock. Indeed all the larger breeding companies in this sector supply products around the world. This situation has evolved both because poultry have a high reproductive rate, and because there is intense competition within poultry breeding and the sophisticated breeding programmes needed to survive this competition are costly.

To expand on the first of these factors, in broilers (chickens raised for meat) one million grandparent stock can produce 50 million parent stock which in turn produce 600 million birds for the market. A single breeding company can therefore supply a huge number of farms. Fish are like poultry in this respect, and cattle are not. In layers (chickens raised to lay table eggs) two companies together serve 80-95% of the European market and over 75% of the world market. Smaller companies are active in domestic and niche markets, and one North American company has a foothold in the European market. Three breeding companies dominate the European broiler market and the world market. Two companies have breeding locations in both Europe and the USA.

One large commercial company also produces non-intensive stocks, and the market in this area is growing: in France, 112 million birds bearing the non-intensive certificate, Label Rouge, were sold in 2000. However, although freerange and organic varieties of bird fetch high prices on the market, little of this revenue trickles down to the breeders, who receive only a small percentage of the mark-up.

Pigs

European pig breeders are currently enlarging their European and global markets. Improved methods of freezing semen for export make this possible. Unlike the situation with cattle, where frozen semen can be used without affecting fertilisation adversely, this results in more conceptions per insemination.

AI usage in pigs differs enormously from country to country, ranging from over 90% of inseminations to virtually none. Embryotransplantation is not yet used commercially in pigs. Within Europe, around 2,500 embryos were transferred in vivo during 2000, but most of these were either used in research or stored to preserve biodiversity.

Fish

The genetic improvement of farmed fish has developed rapidly since the mid-1980s, and nowadays over 70% of EU aquacultural production (especially salmon, rainbow trout and turbot) is from selected stocks. Investments during the 1990s lead to the domestication and selective breeding of sea bass, sea bream and African catfish. Trials of other aquatic species, such as cod, halibut and tuna, are also underway.

From the breeder's perspective, fish are rather like poultry. They are relatively small, produce a large-indeed enormous-number of eggs at a single spawning (between 2,000 and 50,000), and are normally reproduced through artificial fertilisation. As a result of these similarities, the fish breeding sector tends to be evolving along the same lines as the poultry breeding sector, with a small number of companies becoming dominant. In aquaculture, unlike poultry, breeding is often integrated with other activities-e.g. with the husbandry, slaughtering, processing and selling of the fish to the supermarkets.

In France, 16 companies with interests in oysters, rainbow trout, brown trout, brook trout, sea bass, sea bream and turbot combine their research and jointly evaluate their breeding programmes.

The reproduction of farmed fish involves eggs and semen being gently stripped from adult fish by hand. The semen is then introduced into water containing the eggs, and fertilisation takes place very much as it would naturally. Like most plants consumed today, farmed fish may have more than two sets of chromosomes.

This does not mean that they are GM fish. It merely means that, as 'sterile triploids', they are unable breed with their wild relatives when, occasionally, they escape. The practice of inducing triploidy is recommended by international organisations such as the FAO, ICEM and NASCO.

5

Molecular Genetics in Transgenic Animals

INTRODUCTION

A transgenic animal is one that carries a foreign gene that has been deliberately inserted into its genome. The foreign gene is constructed using recombinant DNA methodology. In addition to the gene itself, the DNA usually includes other sequences to enable it:

- to be incorporated into the DNA of the host and
- to be expressed correctly by the cells of the host.

Transgenic sheep and goats have been produced that express foreign proteins in their milk.

Transgenic chickens are now able to synthesize human proteins in the "white" of their eggs.

GENETIC CHARACTERISATION OF ANIMAL GENETIC RESOURCES

Animal Genetic Resources Division is engaged in physical characterization through field survey, sustainable utilization and conservation of indigenous livestock and poultry breeds. Systematic field surveys have been undertaken to assess the phenotypic traits, population status of the breeds and potential population groups, besides the socio-economic conditions of the farmers and communities. In the same vein, the production performance of local breeds and their interaction with local ecology have been evaluated and recorded. Based on the information, new strategies have been formulated for improvement and conservation of the breeds in the farmers' conditions. The in situ conservation has been implemented for different species. In addition, the Division is also working in the frontier areas of ex-situ conservation of germplasm for long term preservation.

Developing countries are endowed with the majority of the global domestic animal diversity-landraces, strains or breeds. Some livestock breeds in these countries are in immediate danger of loss through indiscriminate crossbreeding with exotic breeds. The importance of indigenous livestock breeds lies in their adaptation to local biotic and abiotic stresses and to traditional husbandry systems. However, most of these animal genetic resources are still not characterised and boundaries between distinct populations are unclear. In such cases breeds are defined on the basis of subjective data and information obtained from local communities. Reliance on these criteria as the basis for classification for utilisation and/or conservation may be misleading. Additionally, historical evidence is not always accurate, relying as it often does on subjective judgements. Archival research can reveal much about the original type of a breed or strain but it is molecular genetic evidence which is factual and precise. It is in this sphere that biotechnology has an important role.

Genetic uniqueness of populations is measured by the relative genetic distances of such populations from each other. Polymorphism in gene products such as enzymes, blood group systems and leukocyte antigens which have traditionally been used for measuring genetic distance are being rapidly replaced by polymorphism at the level of DNA, both nuclear and mitochondrial as a source of information for the estimation of genetic distances. The first DNA polymorphism to be used widely for genome characterisation and analysis were the restriction fragment length polymorphism (RFLP) which detect variations ranging from gross rearrangements to single base changes.

Minisatellites sequences of 60 or so bases repeated many hundreds or thousands of times at one unique locus within the genome have been used to generate DNA fingerprints typical of individuals within species. Microsatellites repeats of simple sequences, the commonest being dinucleotide repeats are abundant in genomes of all higher organisms, including livestock. Polymorphism of microsatellites takes the form of variation in the number of repeats at any given locus and is generally revealed as fragment length variation in the products of polymerase chain reaction (PCR) amplification of genomic DNA using primers flanking the chosen repeat sequence and specific for a given locus. Ease of identification and of sequence determination and need for only small amounts of DNA, are some of the advantages of microsatellites. Additionally, because microsatellite polymorphism can be described numerically, they lend themselves to computerised data handling and analyses. Microsatellites can be used in non-PCR systems in a way similar to minisatellite probes.

Randomly amplified polymorphic DNA (RAPD) has been extensively used for genetic characterisation of a wide range of organisms. The technique uses short (up to 10 bases) primers to amplify nuclear DNA in the PCR. The procedure does not require knowledge of the sequence of DNA under study; primers are designed randomly. The basis of the polymorphism detected by this method is that products are either generated in PCR or not.

Complete sequencing of the genome is the ultimate form of genetic characterisation. Sequencing has traditionally been expensive and laborious, but with the advent of automated sequencing this is changing rapidly. However, sequencing is unlikely to be used as a technique of choice for genetic characterisation.

Nei and Takezaki reviewed statistical methods for estimating genetic distances and for constructing phylogenetic trees from DNA sequence data and concluded that different analytical methods may produce different results. Teale *et al.* commented on what considerations to be made in using DNA polymorphism data for genetic distance estimation and cautioned that great care has to be taken in selecting characterisation methods and in interpreting the resulting data. While recognising the importance of the uniparental mode of inheritance of mitochondrial DNA in detecting underlying population structure not discernible from analyses of nuclear DNA, Loftus *et al.* concluded that mitochondrial DNA analysis may not be sufficient to resolve breed differences within Africa. MacHugh *et al.* suggested that microsatellite polymorphism may be more suitable when trying to discriminate between closely related populations. Regardless of which method is used, the ultimate goal in genetic characterisation for conservation is to obtain a measure of available diversity.

Conservation of Animal Genetic Resources

The terms conservation, preservation, ex situ and in situ are used here according to the definition given by FAO. There are several ways, differing in efficiency, technical feasibility and costs, to conserve animal genetic resources. Developing and utilising a genetic resource is considered the most rational conservation strategy. However, there are cases where ex-situ approaches are the only alternatives. *Ex-situ* approaches include: maintenance of small populations in domestic animal zoos; cryopreservation of semen (and ova); cryopreservation of embryos; and some combinations of these. Brem *et al.* reviewed biotechnologies for *ex-situ* conservation.

Cryopreservation of gametes, embryos or DNA segments can be quite an effective and safe approach for breeds or strains whose populations are too small to be conserved by any other means. The safety of these methods has been demonstrated by background irradiation studies. For example, studies based on irradiation of mouse embryos exposed to the equivalent of hundreds of years of background mutation showed no detectable damage. Regeneration of offspring following transfer of frozen-thawed embryos has been successful for all major domestic species, except the buffalo. In cattle, the transfer of frozen-thawed embryos is now a commercial practice and embryo survival rate after thawing can be as high as 80% with a pregnancy rate of about 50%. Cryopreservation of oocytes followed by successful fertilisation and live births have been achieved in the mouse. Cryopreserved bovine oocytes have been successfully matured and fertilised in vitro and zygotes developed to blastocyst stage. These trends strongly suggest that long-term cryopreservation of mammalian oocytes is possible.

Respective pregnancy rates of 58 and 50% for fresh and frozen-thawed *in vitro* produced embryos have been reported. Also, calves have been produced from transfer of both split and frozen-thawed *in vitro* produced embryos.

Economic aspects of genetic conservation in farm animals has been assessed by Brem *et al.*. The study concluded that costs of *ex situ* live animal conservation was moderate to high while costs of long-term cryopreservation of gametes were low.

Development in genetic engineering, cryobiology, cell biology and embryology will provide techniques that may enhance our ability to preserve germplasm *in vitro*. Techniques such as transfer of DNA within and between species and the production of viable transgenic animals are far from practical application. However, biotechnology will certainly contribute newer and cheaper methods for preservation such as storage of catalogued DNA. At present, other than live animal and embryo preservation, the other techniques do not allow preservation of genomes in a form which can be reactivated *in toto* at a later stage, but they permit the preservation of individual genes or gene combinations for possible future regeneration.

Conservation of indigenous animal genetic resources should be one of the priority livestock development activities for developing countries. The critical importance of these resources to their owners in developing countries need not be emphasised. Their importance to developed countries is also becoming evident as indicated by the increasing importation of tropical germplasm by these countries. It is highly likely that these resources will become of increasing importance to the industrialised countries either as sources of unique genes or when environmental concerns necessitate change in production systems. Developed countries should, thus, assist in the conservation and development of these resources. Technology for cryopreservation of semen and embryo is sufficiently developed to be applied in developing countries. What is missing is financial support to implement conservation programmes. Such support has been provided for world-wide conservation activities for plant germplasm. There is also a strong case for support of animal genetic resources conservation.

TRANSGENIC MAMMALS

For the first time in 1982, there appeared a report on the transfer of human growth hormone gene of rat fused to the promoter of mouse metallothioneine I gene. It was done by microinjection method. As a result of presence of a novel gene, there has been a drastic increase in body weight of mice. Since then a large number of transgenic mammals and other animals have been produced such as cow, pig, rabbit, goats, sheep, fish, etc. The purpose of production of transgenic animals has been to produce more protein in milk and meat, disease resistance, leaner meat, good quality wool and more specifically improvement in genetic traits. They are also used as bioreactor for molecular farming.

Transgenic Fish

Fish are the important source of fat and proteins for humans and the delicious diets are prepared in certain societies. Therefore, demand of good quality fish is increasing gradually. Gene transfer in embryos of several species of fish such as medaka fish, salmon, carp, zebrafish, goldfish, trout and cattlefish has been successfully achieved. The novel desired gene is introduced into early embryo through microinjection because pronuclei are not easily visible. In some fish transgene is microinjected into the nucleus of oocyte. The introduced gene replicates at the time of development of embryo. In most of the fish fertilization is accomplished out side in water. Therefore, embryo of different developmental stages can be collected easily. In those cases where microinjection technique is not successful, transgenes are incorporated into embryos through electroporation technique.

Inheritance of transgene occurs in the Mendalian way. Stuart *et al*. found that spermatozoa of transgenic trout and zebrafish containing foreign DNA transmitted the transgene to the next generation. Normal oocyte fertilized with sperms of transgene fish gave rise to offsprings of which 10-50 per cent of Fl offsprings contained foreign genes. The number of transgenic fish increased in F2 generation.

Animal Bioreactors and Molecular Farming

Transgenic animals are used as bioreactor for mass production of drugs and proteins called molecular farming (pharming). It is a new industry. The Transgenic Scien (an American Company) has produced transgenic mice which secreted in milk about 0.5 grams/liter of hGH. Due to expression of transgene, this mice hormone has no adverse effect.

Adverse effect of cattle and pig derived human hormones have been observed. In addition, the pigs that secrete extra hormone, hGH are sterile. This company has started scaling up of production of hGH through transgenic rabbit because of high concentration of protein in milk and short gestation period of rabbit.

A transgenic lamb has been produced by the Institute of Animal Physiology and Genetic Research, Edinburgh (UK). This lamb contains alpha-antitrypsin (αAT) gene. In human the deficiency of αAT causes fluid accumulation in lungs resulting in death of patients. The ?AT inhibits the enzyme elastase that digests foreign particles and clears the lungs. In the absence of αAT elastase is not inhibited and, therefore, it digests lung tissue. In lactating milk of transgenic lamb, presence of αAT can be tested. Transgenic sheep can also be used for production of αAT commercially.

TRANSGENIC SHEEP

So far it is not clear why the rate of transgenesis in sheep is very low *i.e.* 0.1 to 0.2 per cent. It needs improvement by regular check through the biotechnological methods (using PCR, etc). Method of production of transgenic sheep is the same as described for transgenic mice). Due to commercial appeal, the Pharmaceutical Proteins Ltd, Cambridge (UK) provided fund to J.P. Simons for the production of transgenic sheep. In 1988, Simons reported first production of transgenic sheep. He first produced two transgenic ewe that consisted of about 10 copies of human antihaemophilic factor IX gene in the form of cDNA. It was fused with about 10.5 kb long β-lactoglobulin (BLG) gene. Moreover, the BLG gene is important for the expression of gene in mammary gland. The ewes secreted human alpha-1 antitrypsin (hα-1AT) *i.e.* human factor IX in milk because the gene had tissue specific expression. Inspite of low expression of transgene, the hα-lAT is active. The transgenic ewes were born in summer 1986. Again they were mated at the end of the year 1986. Single lamb was born from each ewe in 1987 that inherited BLG-factor IX transgene. Due to the presence of both the genes factor IX was secreted in milk.

In 1991, Alan Colman and coworkers at Edinburgh produced five transgenic sheep, four female and one male. The transgene was ovine-α-lactoglobulin promoter fused to hα-1AT gene. The concentration of h α-1AT in milk was recorded to about 35 grams per liter. The biological activity of protein derived from milk was the same as that of plasma derived antitrypsin.

HEREDITARY DNA OF TRANSGENIC ANIMAL

A transgenic animal is an animal whose hereditary DNA has been augmented by addition of DNA from a source other than parental germplasm through recombinant DNA techniques. Transfer of genes or gene constructs allows for the manipulation of individual genes rather than entire genomes.

There has been dramatic advances in gene transfer technology in the last two decades since the first successful transfer was carried out in mice in 1980. The technique has now become routine in the mouse and resulting transgenic mice are able to transmit their transgenes to their offspring thereby allowing a large number of transgenic animals to be produced. Successful production of transgenic livestock has been reported for pigs, sheep, rabbits and cattle. The majority of gene transfer studies in livestock have, however, been carried out in the pig. Although transgenic cattle and sheep have been successfully produced, the procedure is still inefficient in these species.

Transgenesis offers considerable opportunity for advances in medicine and agriculture. In livestock, the ability to insert new genes for such economically important characteristics as fecundity, resistance to or tolerance of other environmental stresses would represent a major breakthrough in the breeding of commercially superior stock. Another opportunity that transgenic technology could provide is in the production of medically important proteins such as insulin and clotting factors in the milk of domestic livestock. The genes coding for these proteins have been identified and the human factor IX construct has been successfully introduced into sheep and expression achieved in sheep milk. Moreover, the founder animal has been shown to be able to transmit the trait to its offspring. To date, the majority of genes transferred into sheep have been growth hormone encoding gene constructs. Unfortunately, in most cases the elevated growth hormone levels have resulted into a clinical diabetes situation leading to an early death of the transgenic sheep. Transgenic sheep have recently been generated which express the visna virus envelope gene.

The first reports of the production of transgenic animals created a lot of excitement among biological scientists. In the field of animal breeding, there were diverse opinions on how the technology might affect livestock genetic improvement programmes. Some believed that it would result in total reorganisation of conventional animal breeding theory while others considered the technology as an extension of current animal breeding procedures which, by broadening the gene pool, would make new and novel genotypes available for selection. Application of the technology in animal improvement is still far from being achieved. However, consideration needs to be given to its potential role in this field. Smith *et al*. presented a comprehensive evaluation of strategies for developing, testing, breeding and disseminating transgenic livestock in the context of quantitative improvement of economic traits. An important contribution of transgenic technology is in the area of basic research to study the role of genes in the control of physiological processes. The understanding of the molecular control of life processes has important implications for both medicine and agriculture. For example, the generation (through mutation of an endogenous gene) of an organism which lacks a specific gene is a powerful tool to investigate the function of the gene product. This type of genetic analysis has been facilitated by the availability of in vitro cultures of embryonic stem cells from mice. Recent advances in *in vitro* technology (*in vitro* fertilisation and maturation) will increase the number of zygotes available for gene transfer purposes. This, plus the utilisation of embryonic stem cell and primodial germ cell technologies should enhance the efficiency of gene transfer in cattle and sheep considerably.

APPLICATION OF MOLECULAR GENETICS

Methods have been developed to isolate animal genes and to characterize them. The first approach is to identify DNA sequence associated with economic trait loci.

The second step is to incorporate the identified gene into three maps:

1. A physical map of chromosome (where DNA sequences are assigned to specific sites on specific chromosomes),
2. A linkage map (where linkage of different genes and markers is assigned in the same chromosome),
3. A genetic map (where inheritance of economic traits is corrected with the inheritance of genes and markers).

The unknown and novel genes associated with genetic diseases can also be identified through their expression. Then the protein sequences are identified through amino acid sequencing or injecting the protein into suitable animal (*e.g.* mouse, rabbit, chicken, etc.) to raise the antibodies.

The animal production industry is gradually reaching towards revolution due to the development of detailed linkage and genetic map, more knowledge of expression and regulation of genes and techniques for large propagation. The application of molecular genetics in animals currently being used are: breeding selected traits into livestock, animal cell culture, and production of transgenic animals.

Selected Traits and their Breeding into Livestocks

There are several genetic traits known which can be bred into livestock if required, for example disease resistant traits. Some traits are determined from the coordinated expression of many genes but not single gene. The discipline of quantitative genetics unravels the total number of genes contributing to a trait, contribution of individual trait and its location on the chromosomes. Altogether they are termed as quantitative trait loci (QTLs), therefore, herds are surveyed for reference animals that have been bred for a broad range of phenotypic variations.

Diagnosis, Elimination and Breeding Strategies of Genetic Diseases

An increase in homozygosity of recessive genes and the frequency of specific alleles at QTLs has been resulted due to isolation of domestic animals in small breeding group. Consequently most breeds of livestocks comprises of variant genes which are harmful or lethal at homozygous stage (when it contain two variant genes). Such allele may be found in 15 per cent cases with 5 per cent homozygotes. Several variant genes responsible for genetic diseases have been identified. Diagnostic tests are available which permit animals as normal, carrier (*i.e.* heterozygous with one normal allele or one variant allele) or affected (*i.e.* homozygous with two variant alleles). Now PCR is used to amplify the region of genes which has been affected. Thus amplified DNA fragement is used to identify the variants. It is well known fact that genetic disease occurs when a gene malfunctions and enzyme is not expressed. When both alleles are defective in homozygote these results in phenotypic defects. The carriers (heterozygote) are phenotypically normal. Therefore, the genetic defects can be diagnosed by genetic tests and rectified biotechnologically during pregnancies that affect foetus.

The second approach for elimination of defect from a herd is the use of normal (homozygous) animals for mating. Use of MOET with embryo biopsy and splitting on the best animals is the other method of spread of genetic disease and curing them. Pre-identified carrier or affected embryos should not be transferred (this aspect has been discussed earlier).

Application of Molecular Markers in Improvement of Livestock. During the last two decades the progress in recombinant DNA technology and gene cloning has brought in revolutionary changes in the field of basic as well as applied genetics. Several new approaches for genome analysis have been made. Now it is possible to uncover a large number of genetic polymorphism at the DNA sequence level and to use them as markers for evolution of genetic basis for the observed phenotypic variability.

Usually a marker is considered as a constituent that determines the function of construction. Variations occurring at different levels *i.e.* at the morphological, chromosomal, biochemical or DNA level can serve as genetic markers. Thus genetic markers can be defined as any stable and inheritable variation that can be measured or detected by a suitable method, and that can be used to detect the presence of a specific genotype or phenotype other than itself. The markers revealing variations at DNA level are referred to as 'molecular markers'. So far an unlimited number of molecular markers is known since the first demonstration of DNA level polymorphism in 1974, which is called restriction fragment length polymorphism (RFLP). The molecular markers are classified into two broad categories, the hybridization-based markers and the PCR-based markers.

The Hybridization-Based Markers

This includes the traditional RFLP analysis. During RFLP analysis well labeled probes for important genes (*e.g.* cDNA or genomic sequence) are hybridized onto filter membrane containing restriction enzyme digested DNA. Then these are separated by gel electrophoresis and subsequently transferred onto these filters by Southern blotting.

Thereafter, the polymorphisms are observed as hybridization bands. The individuals that carry different allelic variants for a locus will show different banding pattern.

Hybridization can also be carried out with the probes (*e.g.* genomic or synthetic oligonucleotide) for the different families of hypervariable repetitive DNA sequences such as minisatellite, simple repeats, variable number of tandem repeats (VNTR) and microsatellite to reveal highly polymorphic DNA fingerprinting pattern.

The PCR-Based Markers

There is no need of probe-hybridization step. The PCR-based markers have lead to discovery of several useful methods which are easy to screen. On the basis of types of primers (*i.e.* primers of specific sequences targeted to particular region of genome or primers of arbitrary sequences) used for PCR.

These markers further can be subdivided into two groups, the sequence-targeted PCR assay, and the arbitrary PCR assay:

1. The sequence-targeted PCR assay. In this assay system, a particular fragment of interest is amplified using a pair of sequence-specific primers. In this category, PCR-RFLP or cleaved amplified polymorphic sequence (CAPS) analysis is a useful technique for screening of sequence variations that give rise to the polymorphic restriction enzyme (RE) sites. A specific region of DNA encompassing the polymorphic RE sites is amplified. The amplified DNA fragment is digested with respective RE. The variations in sequence are screened by several approaches.
2. The arbitrary PCR assay. In this assay randomly designed primer is used to amplify a set of anonymous polymorphic DNA fragments. When primer is short, there is high probability of priming taking place at several sites in genome that are located within amplifiable distance and are in inverted orientation. Polymorphism detected by using this method 'is called randomly amplified polymorphic DNA (RAPD). Based on this principle several techniques have been developed which differ in number and length of primer used, stringency of PCR conditions and the method of fragment separation and detection.

Properties of Molecular Markers

In genetic analysis, many types of markers *viz.*, morphological, chromosomal, biochemical and molecular markers are used. Morphological (*e.g.* pigmentation and other features) and chromosomal (*e.g.* structural and numerical variations) markers usually show low degree of polymorphism, therefore, they are not very useful. Biochemical markers have been tried out extensively but have not been found encouraging as they are sex linked, age dependent and influenced by the environment. The molecular markers capable of detecting the genetic variations at the DNA sequence level, have removed the limitations. They possess unique genetic properties that make them more successful than the genetic markers. They are numerous, distributed on genome, follow typical Mendalian inheritance and are multiallelic giving heterozygosity of more than 70 per cent and unaffected by environmental factors.

For genetic analysis, molecular markers offer several advantages.

Mitra et al. (1999) have discussed the several advantages of molecular markers:

- The DNA samples can easily be isolated from blood, tissues *e.g.* sperms, hair follicle as well as archival preparations,
- The DNA samples can be stored for a longer time and readily be exchanged between the laboratories,
- The analysis of DNA can be carried out at an early age or even at the embryonic stage, irrespective of sex,
- Once the DNA is transferred onto a solid support *e.g.* filter membrane, it can be repeatedly hybridized with the different probes and heterologous probes and in vitro synthesized oligonucleotide probes can also be used, and
- The PCR-based methods can be subjected to automation.

Application of Molecular Markers

Polymorphism observed at the DNA sequence level has been playing a major role in human genetics for gene mapping, pre-and post-natal diagnosis of genetic diseases, and anthropological and molecular evolution studies. Similar approach for exploitation of DNA polymorphism as genetic markers in the field of animal genetics and breeding has opened vistas in livestock improvement programmes. Much interest has been generated in determining variability at the DNA sequence level of different livestock species, and in their assessment whether these variations can be exploited efficiently in conventional as well as transgenic breeding programmes. Molecular markers can play important role in livestock improvement through conventional breeding strategies. The various possible applications of molecular markers are: short-range (immediate) applications and the long-range applications.

Transgenic Breeding Strategies

The current breeding strategies of livestock largely rely on the principle of selective breeding. In this method genetic improvement is brought about by increasing the frequency of advantageous alleles of many loci. The actual loci are rarely identified. In these methods genes cannot be moved from distant sources like different species or genera due to reproductive barrier.

The recent development in molecular biology has given rise to new technology called transgenesis, which has removed the breeding barriers between different species or genera.

Transgenesis has opened up many vistas in understanding behaviour and expression of a gene. It has made possible to alter the gene structure and modify its function. There are many applications of transgenesis, but the most convincing one is the development of transgenic dairy animals for the production of pharmaceutical proteins in milk, and animals with altered milk composition.

The transgenesis first starts with identification of the genes of interest. In this context, molecular markers can serve as reference point for mapping relevant genes. After successful production of transgenic animals, appropriate breeding methods could be followed for multiplication of transgenic herd. Molecular markers can also be used for identification of the animals carrying the transgenes. Most of the QTL are polygenic in nature and transgenesis presently single gene traits are being manipulated. The technology holds future promises in moving polygenic QTL across the breeding barriers of animals. However, it is expected that molecular markers wili serve as a potential tool to geneticists and breeders to evaluate the existing germplasm, and to manipulate it to create animals of desired traits as demanded by the society.

FOOD OF TRANSGENIC IN ANIMAL BIOTECHNOLOGY

Transgenosis technology is a kind of modern technology in molecular biology, which is used to transfer genes from one species into another so as to reconstruct the genetic materials of the receiving species for the improvement of its properties, quality of nutrition in line with the need of human beings. The transgenic species as immediate food and food processed from transgenic species are called transgenic food. In other words, transgenic food are those directly made from or processed from the species (animals, plants and microorganisms, etc.) which can produce substances possessing highly effective expressions, such as polypeptide and protein, after one or several types of exogenous genes are transferred into it through the means of genetic engineering.

SCIENCE, INSTITUTIONS, SOCIETY AND POLICIES

There are many scientific uncertainties relating to animal biotechnology. Current understanding is considered to be inexact, uncertain and insufficient as regards technique, methodology and epistemology. There are too many factors, variables and interactions to accurately assess the level of risk. These interactions are not always known or understood. Genetic engineering has unexpected secondary consequences and the aggregate impact of these consequences on intended transgene impact cannot be anticipated. Similarly, genes have pleiotropic effects that are impossible to foresee. It is therefore clearly impossible to envisage all the potential risks. Some of these may not be identified and therefore not studied. On the other hand, analysis of a non-existent risk is a waste of time and resources, and can result in unnecessary costly management measures. In addition, estimating potential risk by using inherently biased methodology must necessarily lead to error. Regulatory bodies make their decisions on the basis of science-based risk assessment.

Scientific uncertainties should therefore be priority research areas in both the private and the public sector. A process of consultation with all stakeholders is required to establish agreed and validated risk assessment protocols and to share results and experiences. There are at least four direct consequences of existing scientific uncertainties: application of the precautionary principle in certain national regulations; extreme precaution in risk analysis; stated need for post-approval monitoring of transgenic animals; and the need for further research.

Policy Context, National and International Agencies

Animal biotechnology raises a number of policy problems. Which standards apply to biotechnology products? A major problem is that the risks concern a variety of areas: human and animal health, animal welfare and the environment. How can these different areas be integrated into the decision-making and regulatory process? The welfare of transgenic animals is usually covered by legislation on animal welfare in general. Is specific regulation on the welfare of transgenic animals required? Another problem is the absence of regulation on transgenic insects.

There is clear need to clarify the scope and limits of different regulatory and control agencies, and national and international organizations as regards the different areas that affected by biotechnology. Should their remit be extended beyond public health and environment to include social and economic aspects, given that these are also relevant? Many different agencies are involved: those responsible for the environment (field releases and trials), food and human health, imports and exports, and animal protection and health. It is important to define individual responsibilities, legitimacy and role.

Transgenic animals and derived food products are governed by regulation on novel foods (Novel Food Regulation introduced in 1997 by the European Commission) in Europe and (Novel Food Regulation part of the Food and Drugs Act of October 1999) in Canada. For Canada, the concept of novel food includes any food from plants, animals or micro-organisms that has been genetically modified in such a way that new traits have emerged, or are no longer observed, or are observed at unusual levels. In Canada, protocols are drawn up by Health Canada to assess the safety of novel food. In the USA, the US Federal Drug and Cosmetics Act has been amended to include GM food.

To summarize, approaches differ in many respects:

- *Basic legislation*: Some countries (Canada, USA) have incorporated GMO legislation into existing legislation; others (Australia, Japan) have issued specific legislation for biotechnology, resulting in a mass of sometimes extremely complex regulations.
- *Transparency of decision-making*: Some countries do not consider public opinion (Canada) and only provide a summary of information once a decision has been made, without prior notification. Conversely, others announce their intentions, consult public opinion and issue reports that summarize positions and decisions (EU, USA).
- *Approach to risk assessment and in particular to the principle of precaution*: Some countries seek to eliminate all risk and only decide once every aspect has been assessed; others decide on the basis of existing rather than potential risks.
- *Conceptual approach*: For some, the risk to be assessed concerns the product (Canada), while for others it concerns biotechnology in general (Japan, EU, Australia)
- *Independence of decision-making*: Are all links between governments, scientists, industry and expert committees legitimate? In the USA, there is sometimes a close link between decision-makers and assessors.

Good practices often advocated include greater transparency and public involvement, standardized risk assessment concepts, uniform understanding of the principle of substantial equivalence, and distinction between technique and product. According to the British Better Regulation Task Force, regulation should be transparent, accessible, targeted, complete and proportionate to cost and risk.

Convergence of Policy, Science, Ethics and Public Opinion

Concerns that are not science-based should not guide scientific examination. However, such concerns do inherently influence the scientists' mindset, the way they address questions to policy-makers and their presentation of scientific response to public health and environmental concerns. At the same time, government agencies are influenced in their priorities and decisions by the issues raised by public opinion.

The Consumers' Perspective

Recent events regarding the application of modern biotechnology to animals and food production have received considerable attention and raised serious public concern. The introduction of new characteristics into food that differ significantly from those of traditional foods could accentuate these concerns in the future. It therefore seems necessary to keep the public informed of safety assessment procedures applied to novel foods. Another difficulty is the differing consumer needs and expectations in the world regarding food quantity and quality, and the differing nutritional values and cultures. These factors need to be taken into account when examining food production, trade and recent events such as crop enhancement and functional foods. Food safety is a universal concern of consumers who need reassurance that marketed transgenic products have been properly tested for deficiencies and checked for safety. The five eurobarometers conducted in the European Union since 1991 (the last in 2002) give a certain idea of how the European public perceives GMOs.

In many countries this perception is in fact increasingly negative, with concerns relating to:

- The unexpected effects and consequences of gmos on human health and environment;
- The potential irreversibility of negative effects;
- The ethical acceptability of gmos;
- The usefulness of gmos;
- Animal welfare;
- Relational imbalances between producers, farmers and consumers;
- The equality of peoples in the face of biotechnology.

There is a clear distinction in the way GMOs are perceived according to their intended application. On the one hand, we have biomedical applications which are considered useful, morally acceptable and are encouraged, while, on the other, we have agrifood applications which are considered redundant and hazardous to society. Public reaction varies according to familiarity and preoccupation with the subject. However, some observers report a difference between what the public says and what the public does or is. People *think as citizens and not as consumers*. The perception of risk differs between the experts and the public. The way the experts estimate risk does not appear to coincide with the public's perception of risk. There appears to be a lack of trust in regulatory bodies and scientific experts, which in part explains the increasing consultation of public opinion, the greater emphasis on transparency and the release of public information. Assessors have identified three categories of public opinion: unconditional supporters, those who accept the risks and opponents.

Acceptance of GMOs is conditioned by four factors: material values, optimism, confidence and preoccupation with issues linked to GMOs and derived products. Survey responses vary according to connotations underlying related terms, such as "genetic" and "biotechnology", so depend directly on the social and scientific climate at the time of the survey. Consumers believe they have a right to be informed on the products they purchase. However, the question of whether the labelling of transgenic foods is the most appropriate and realistic way of enabling consumers to make enlightened choices is a matter of fierce debate in many countries. The Codex Alimentarius Commission is currently examining this issue.

Several governments have adopted policies and procedures on GMO labelling, and significant differences exist. Provisions for labelling from production to consumption can pose insurmountable problems for countries with limited capacity but anxious to enter international markets. Public opinion in the USA appears to be less animated, and opinions seem stronger on genetically modified animals than on genetically modified crops. The American people appear to be aware of and encourage the potential benefits of biotechnology, with one survey revealing that 62% believe biotechnology will produce additional benefits over the next five years. Such benefits include improved quality, taste and diversity of food, improved human health and nutrition, reduced use of pesticides and insecticides, lower food costs and improved yields. The current use of GM crops does not seem to be a problem for the US population. Do they feel the same way about GM animals? Survey data reveal inadequate methodology, with the use of closed questions and an interpretation of results that needs to be shorn of the deep-seated mental prejudices of both assessors and the public. Questionnaires often focus on GM plants, while it might be illuminating to distinguish between GM plants and GM animals to gauge the level of public concern about genetically modified animals.

Communication between the Public, Assessors and Decision-makers

The OECD countries and intergovernmental organizations are seeking new ways of sharing their experience. They are seeking to boost the dissemination of information and real consumer understanding of the safety issue. Some countries have adopted measures for the public exchange of information on the safety assessment of GM foods. The public are invited to comment on the reports and proceedings of safety assessment bodies.

Some regulatory bodies actively consult the public on food safety and regulations. Some make all information supporting an application available to the public, except confidential commercial information. The Internet is increasingly used to inform the public of safety assessment and approval procedures, and is a good source of information on approved foods and food products.

Some countries are looking into the possibility of using the Internet to make application details accessible to the widest possible audience, in order to make the assessment process as transparent and understood as possible. The OECD's on-line BioTrack site is a useful source of information on regulations in member countries, with information on ministries and agencies, and details of laws, regulations and guidelines. It also has two important databases: one on products that have been marketed; the other on GM field tests carried out in OECD countries. Other approaches adopted with huge success in selected countries include the appointment of consumer representatives to safety assessment committees. Consensus conferences have also been attempted with varying degrees of success.

Ethical Considerations

Up until now, the idea of food products from transgenic animals has not been well received by consumers. Surveys have constantly shown that the public is better disposed towards transgenic plants than transgenic animals. There is greater reluctance towards trials involving animals and changes in animal make-up and therefore broader possible implications. Various cultures and religions limit or forbid the consumption of certain foods of animal origin. The public seems however to be more disposed to consume by ingestion or injection pharmaceutical products derived from transgenic animals. Various ethical issues have been raised.

Distribution of Benefits

One preoccupation is the possible impact of biotechnology on agricultural structure, especially the potential elimination of small livestock farms in the face of the cost of producing transgenic animals. People are worried about the impact of a dismantled agricultural society on the environment, sustainable agriculture and quality of life. The public points to the concentration of production and its direct consequences on markets and consumers, and is wary of the dependence of livestock farmers and food producers on multinationals. There is mental conflict between the law of the market and the traditions, cultural identities and needs of society.

Nutrition and Food Safety

There seems to be general consensus among scientists that the consumption of food derived from transgenic animals is no more hazardous than the consumption of conventional products. However, the public thinks differently. Some people refer to specific diets based on cultural or religious considerations. For example, the consumption of an animal or a food from a transgenic animal that contains proteins or genes of animals whose consumption is forbidden by culture or religion could pose a problem. Another point often stressed is the social divide from different levels of access to biotechnology in developing and developed countries.

Respect for Life and the Unnatural Nature of GMOs

Some people do not believe that technical and economic pressures justify the commercialization of life and the patenting of living organisms. Overstepping of the species barrier and the large-scale industrial production of genetic animals are seen as contrary to nature. This argument is frequently refuted by biologists who stress that the continuous natural evolution or the human selection of animal genotypes has favoured certain genes and eliminated others. Finally, biotechnology and conventional livestock practices might seem to pose similar problems, in that both seek maximized productivity and both cause animal dysfunctioning and welfare and health problems. However, such problems seem to be magnified (some would say disproportionately) in the context of biotechnology. Animal biotechnology is a recent science and therefore difficult to view in the light of experience. Many uncertainties and many doubts exist. Information currently available will have to be constantly updated with data and results of analyses and trials of transgenic animals and derived food products, so that techniques, regulations and public opinion can evolve and adapt; so that doubts can be clarified and problems resolved; and so that agreed protocols of analysis and international standards can be determined. Given the rapid progress being made and the prospects of animal biotechnology, it is almost certain that improved techniques will boost benefits. However, the trade-off between risk, cost and benefit will always have to be examined. Ongoing oversight, greater transparency of results from scientific trials, greater transparency of regulatory and decision-making processes and closer cooperation among all stakeholders are required for the safety evaluation of products derived from transgenic animals.

TRANSGENIC ANIMALS AND RELIGIOUS FOOD LAWS

Transgenic animals pose problems for religions that restrict the foods that their believers can eat, since they may produce animals that appear to be one species, but contain some elements of a forbidden species.

Ethical Issues of Transgenic Animals

Transgenic animals raise several particular moral issues (quite apart from any damage they might do to the environment): Are animals that combine species an unethical alteration of the natural order of the universe? Is it unethical to modify an animal's genetic make-up for a specific purpose, without knowing in advance if there will be any side-effects that will cause suffering to the animal? Does 'creating' animals by genetic engineering amount to treat the animals entirely as commodities? Is it unethical to create 'diseased' animals that are very likely to suffer? Suffering may last for a long time in these animals as researchers want to conduct long-term investigations into the development of diseases.

SAFETY OF FOOD OF TRANSGENIC ANIMALS AND DERIVED PRODUCTS

Concern about the quality and safety of food products has been growing since the early 1980s. Many hazards have been identified, studied and handled by instituting national regulations and international standards and by setting up regulatory and inspection systems in processing plants, livestock farms and slaughterhouses.

These hazards can be biological (bacteria, parasites, viruses, prions), physical (metals, etc.) and chemical (residues of pesticides or veterinary drugs, toxic substances, etc.). New hazards are now perceived in relation to the use of transgenic animals. Transgenic animals include those obtained through nuclear transfer, those with improved production traits and those developed for biomedical purposes. Animals for xenotransplantation are barred from the food chain by national regulations.

Products Derived from Cloned Animals

In 2002, some 1400 Holstein cattle produced by blastomere nuclear transfer (not somatic cells) were raised in the US. They and their subproducts (animal and milk) are marketed and enter the food chain. No result of analysis of products derived from these transgenic animals is so far available, nor is there any information on abnormalities of gene expression in these adult cloned animals or analytical information on the composition of milk or meat derived from them or their offspring. The animals are considered perfectly safe and no apparent ill effects have been noted among consumers. However, caution and risk analysis are recommended for such derived products, especially in the light of the problem of reprogramming the inserted nucleus.

The safety of food from offspring of cloned animals is a controversial subject. Some see no risk as they originate from natural mating, while others point out that the offspring of a cloned animal has one half of the latter's genome. The cloning of animals from somatic cells is a recent technique, so it is difficult to make a firm decision on the safety of related milk or meat.

Genetic variations appearing with nucleus differentiation and possibly causing modified genome expression are not known. The findings of studies on products of cloned animals will no doubt be available soon.

Allergenicity and Hypersensibility

Most food allergens are proteic, well known and identified. As regards the safety of new proteins in foods derived from transgenic animals, the possibility that a new protein could cause an allergic reaction among certain individuals needs to be assessed. FAO has published an approach for assessing the potential allergenicity of new proteins extracted from foods derived from transgenic animals.

If the new protein is produced from a transgene that is known to have allergenic potential, or if the animoacid sequence is similar to that of a known allergenic protein, then the allergenic potential of the new protein can be tested with the serum of individuals with known food allergies. The problem arises for proteins with unknown allergenic potential which cannot be easily assessed, as there is no existing human immunoserum. The problem is further compounded when the new protein derives from an existing protein that has no known allergenic potential.

Thus, assessing the potential allergenicity of a protein from a transgene remains one of the major difficulties in the overall safety assessment of transgenic foods. Assessment of allergenicity requires an understanding of several factors, including the source of the transgene protein, its level of expression, its chemical and physical properties, and its similarities to any known allergen.

All these factors have to be considered together to have an idea of the potential allergenicity of the new protein.

The absence of appropriate and effective methods of analysis is a major problem, as is the feasibility of performing adequate assessments for an increasing number of transgenic products. In this connection, WHO and FAO are working on the development of a standardized protocol for the assessment of allergenicity.

The animal model for allergenicity tests of products derived from transgenic animals is recognized as effective. In vivo allergenicity tests among humans would be the best option but raises ethical problems. Post-market surveillance of a product is also advocated, but modalities for doing this are difficult to determine.

Other Bioactive Compounds

Some transgenic animals are designed to express biopharmaceuticals. That these products could retain their bioactivity in the organism after consumption of the food is a safety concern. All proteins and polypeptides are normally broken down in the digestive tract, thus destroying their bioactivity. Only small peptides pass the barrier of the digestive epithelium. The problem remains, however, for people whose digestive epithelium has been compromised by disease or injury. Similarly, the digestive epithelium of young children is permeable to proteins for a period of a few weeks to a few months, during which these proteins could enter the bloodstream. This poses a real problem for bioactive peptides contained the milk. The peptides or proteins could have a negative impact on the digestive system. The production of bactericidal proteins in milk to reduce cattle susceptibility to mastitis from *Staphylococcus aureus* or to control bacterial disease in fish could upset the bacteriological equilibrium of the intestinal flora of consumers.

It could also lead to pathogens resistant to these antimicrobials. Many transgenic fish express a growth hormone gene, which is usually of fish origin. Can this growth hormone cause problems among consumers? The FDA reports that studies have shown that non-human growth hormones, whether of bovine, ovine, porcine or whale origin, are not biologically active among humans. Nor are digested peptide fragments or insulin-like growth factors I secreted in response to GH administration. Moreover, the IGF-I is destroyed by heat along with the active growth hormones. In the (unlikely, given their age) case of young children being fed transgenic fish or milk containing GH, this would be destroyed during heating. The problem however remains for individuals whose digestive tract has been impaired by disease, digestive ulcer or age and which therefore absorbs proteins that have not been broken down and introduces them into the organism. So the risk varies according to the individual. For a susceptible individual, the hazard could have severe consequences, while for a healthy individual of any age, the risk is thought to be low to moderate.

Toxicity

Because proteins are partly broken down in the digestive tract, their direct toxicity is only a mild food safety concern. Moreover, any direct toxicity of proteins would normally be detected during pre-market assessment. The risk is therefore minimal. The problem arises in connection with unintended and unanticipated effects of products derived from GM animals, linked not only to the proteins that are directly expressed by a transgene but also to any unintended by-products. For example, the insertion of a transgene could disturb a specific metabolic process and cause the expression of metabolites with some toxicity. What we need to know is whether there are effective methods of analysis and screening that can detect a toxic substance or a change in metabolism that introduces toxicity. Whole foods can have a tiny concentration of proteins expressed by a transgene or arising as a by-product of the transgene. Such toxicity can be very difficult to detect given the sensitivity limits imposed by laboratory animal numbers, the small quantities of substances tested in whole foods, and the possibility that any negative effect will seldom occur.

Methods include:

- Analysis of the structural/functional relationship of the substance and of its physical and chemical properties and comparison with existing databases of toxic agents to see if the molecule has similarities with a known toxic agent.
- *In vitro method:* Testing the stability of molecules under different conditions or the ability of the molecule to act on other molecules with specific functions. Few *in vitro* tests have been validated and extrapolated *in vivo* as this is difficult, for the method is still fraught with uncertainties.
- *Animal model*: The best indicator of possible toxicity. However, extrapolation remains uncertain, hence the use of safety factors. This model serves to determine acute, subchronic or chronic toxicity. The problem is that the dietary regime needs to be balanced if parallel side effects are not to be caused (high possibility of bias). A precise diet, specific doses and clear protocols are required.

Institutions, Scientific Expertise and International Standards

International agencies have been charged with establishing international standards for GMOs and derived products. Codex is developing principles for risk analysis applied to foods from GMOs. The first principle is to have a pre-market assessment involving case-by-case study of the direct and indirect consequences of inserting a new gene (toxicity, allergenicity, properties of compounds, stability of genome, nutritional effects and other impacts). Codex principles are however for reference only and not binding.

Evaluation of Safety of Foods Derived from Genetically Modified Animals

As concerns examination of the safety of GM foods, the risks associated with genetic engineering are the involuntary modification of the host genetic substance, which can induce changes in food components, including the toxic substances.

Evaluation of the safety of GM foods should therefore include the following procedures:

- Identification of the organism that has been modified and of the source organism(s) of the inserted gene(s). The origin and nature of all the genetic elements that have been introduced into the modified organism need to be identified, including the structural and regulatory sequences and all other parts of the vector sequences. The possibility of horizontal transfer of genes and related consequences also needs to be studied. If the transfer of genes resistant to antibiotics is intended, the possible health implications of concentrations of micro-organisms in the human digestive tract or other parts of the body need to be considered.
- Identification of primary and secondary genetic products, including a description of the characteristics of the inserted gene. Identification of hazards requires an understanding of the expressed inserted genes, the characteristics, concentration and location of expressed food products and the consequences of expression. For each protein resulting from genetic modification, the factors taken into account include possible toxicity and allergenicity from exposure to the expressed protein.
- Evaluation of the safety of new substances anticipated in the foods (*i.e.* proteins, glucides and lipids).
- Assessment of unanticipated effects on the composition of foods, including assessment of changes in the concentration of nutritive or toxic substances occurring naturally, identification of anti-nutritional factor compounds significantly altered in GM foods and assessment of the safety of compounds showing significant modified concentration.
- Evaluation of any toxin produced directly by modification.
- Evaluation of type of treated food, assuming normal manufacturing procedures.
- Evaluation of food consumption patterns, including identification of potential human population consuming GM foods and probable consumption levels, and evaluation of possible consequences if the consumption of modified foods differs from that of equivalent traditional foods.
- Assessment of allergenicity of new foods.

Current approaches to assessing the toxicity of specific proteins are based on internationally approved, acute or chronic testing on laboratory rodents or on fast-growth domesticated species, such as chickens. However, no repeated dose testing with GM foods or food products has yet been published. As regards allergenicity, most scientists now base their evaluations on a decision tree that has been designed to evaluate the potential allergenicity of foods from GM plants, in the light of current knowledge of what is defined as an allergen and how to better identify an allergen.

Concept of Substantial Equivalence

The OECD's principal contribution to the evaluation of safety of foods from genetically modified animals has been its publication *Safety evaluation of foods derived by modern biotechnology: concepts and principles*. According to this report, the most practical way of determining the relative safety of a new food product is to determine whether it is substantially equivalent to its conventional counterpart, where this exists. This concept was initially applied to organisms of terrestrial origin but, with certain qualifications, was then also applied to products of aquatic biotechnology. Substantial equivalence recognizes that the purpose of evaluation is not to guarantee absolute safety, but to determine whether the transgenic food presents more risks than its conventional counterpart, should this exist. It is generally agreed that such evaluation requires an integrated, specific and gradual approach. The factors taken into consideration when a transgenic food is compared to its conventional counterpart are:

- Its identity, source and composition; its characteristics in relation to the "natural" product;
- The effects of processing and cooking;
- The processing stage, the dna itself and the products of protein expression of the introduced dna; its nutritional qualities;
- The effects on its function;
- Its potential toxicity and allergenicity, as well as possible secondary effects;
- Consumption of the transgenic food and the nutritional impact of its introduction.

Interactions between environmental factors and animal genotype also need to be taken into account. These variations can influence the phenotype and composition of the modified animal and need to be studied during the evaluation of equivalence. If the food derived from a GMO is judged to be substantially equivalent to its conventional counterpart, it is then considered to present the same level of risk. Otherwise, further tests are needed. Confirmation of substantial equivalence does not mean that the product of a genetically modified animal is identical to the conventional product. Comparison does not take all elements into account and substantial equivalence only provides assurance that the appropriate components of safety of the product will be present in equivalent quantities.

As the comparative approach links the composition of new foods to existing products with a history of safe use, the consequences of introducing new foods into a diet can be foreseen. The concept means that all similarities between foods from genetically modified animals and conventional foods can be judged to be safe.

Differences noted during comparison are subject to further meticulous examination which includes the testing of traditional nutritive elements and toxicological or immunological testing or long-term studies, depending on the differences noted. Foods from transgenetic animals might not have any conventional equivalent or the differences between these foods and their conventional equivalents may not be sufficiently characteristic. The history of safe use of the conventional product could not therefore be used to assess the safety of the new food under study.

Substantial equivalence cannot be used to determine the safety of foods for which there is no comparator. However, that a novel food does not bear much resemblance to its counterpart does not necessarily make it dangerous. Its safety needs to be determined by using traditional food safety approaches, with supplementary testing on the consequences of modification.

Post-Approval Monitoring of Foods Derived from Transgenic Animals

It is widely accepted that foods and food products need to be thoroughly evaluated before being placed on the market, if consumers, animal health and the environment are to be protected. If the assessment raises safety concerns, the product will not be approved for commercial release. Some countries therefore fail to see any scientific justification for the post-approval monitoring of these products. Other countries, on the other hand, see the monitoring of possible negative or positive impacts of novel foods as a logical follow-up to the initial science-based risk assessment.

They argue that post-approval monitoring should be mandatory to consolidate the initial pre-market risk assessment, which was based on the then state of scientific knowledge. This could apply to any pre-market assessment. It is recognized that it is sometimes difficult to draw clear conclusions from epidemiological studies, especially when applied to food compounds. Post-approval monitoring is advocated as the only feasible way of demonstrating the absence of possible long-term or unexpected consequences of consuming a novel food. Implementing this strategy is not easy, however.

STRATEGIES FOR GENE TRANSFER

Desired foreign genes are transferred into animal cells/embryos via virus, microinjection, targeted gene transfer methods, etc. as discussed below:

Clinical Gene Transfer for Hemophilia

On the basis of considerations discussed above, it is likely that in the case of hemophilia, more than one successful gene-based approach to treatment can be developed. In a sense, this can be viewed as analogous to the situation with antibiotics: multiple drugs are available, and physicians choose the most suitable one on the basis of the organism to be treated and the side effects of the drug.

In the case of hemophilia, the disease state to be treated is the same, but patients differ considerably in comorbid conditions that may influence the choice of treatment. Thus some patients are on antiretroviral drugs and cannot be treated with retroviral or lentiviral regimens, whereas those with liver disease from hepatitis infection may not be good candidates for liver-directed treatment approaches.

At this point, the hemophilia population is probably best served by the continued simultaneous development of multiple approaches; one of the major challenges of the next few years of clinical research in this area will be to define which subgroups of patients can be most safely and effectively treated with which approaches.

If some approaches are much safer and more effective than others, then these will become the benchmark standards against which new strategies will need to be assessed.

Gene transfer strategies are characterized by 3 essential elements: the gene delivery vehicle, or vector; the gene to be transferred, sometimes referred to as the transgene; and a specific target cell, which may determine the route of administration of vector.

Gene transfer strategies are often referred to by a shorthand that lists these 3 key elements, eg, adeno-associated viral (AAV)-F.IX-liver. Another important characteristic of gene transfer is whether it is given ex vivo (outside the body), as is generally done when hematopoietic stem cells are the target, or in vivo, ie, vector is injected directly into the recipient. Both ex vivo and in vivo approaches are currently under investigation as methods of treating hemophilia.

Transfection of Animal Cells/Embryos

The transfected cultured mammalian cells have been used for diagnostics of oncogene (*i.e.* cancerous gene) as well as for gene therapy.

The steps for detection of cancer gene follow:

- Isolation of DNA from tumour cell line,
- DNA fragementation into 30-50 kb long pieces through mechanical shearing,

- Dissolution in phosphate buffer followed by precipitation by adding $CaCl_2$,
- Pouring of this solution onto a layer of mouse 3T3 cells; foci of cells developed,
- Use of transfected cells in detection of cancer causing genes.

The other application of transfected cells is in gene therapy. Genes of desired function are inserted in cultured cells. The later is placed in patient's body to rectify the malfunctioning gene.

Transfer Through Microinjection

Microinjection method has also been developed and variously used in production of transgenic animals. So far gene transfer has been successfully carried out in several classes of animals *viz.*, fish, birds, insects, mammals, etc.

The Targeted Gene Transfer

The other approach is the targeted gene transfer that involves transfer of genes at homologous sites in the host genome. It is done just to replace the wild type of mutant genes. For the first time it was done in bacteria and yeast. In 1985 success has been achieved in human also where human ?-globin gene was transferred into recipient cell through recombination. Targeted gene transfer is possible because the homologous DNA sequences arc present at the targeted site as well as in vector that carries the desired gene of foreign origin.

Gene targeting is also done by using embryonic stem (ES) cells as described earlier. The ES cells are allowed to get transfected by vector containing desirable genes. In transfected cells targeting of gene to specific site by homologous recombination occur.

Thereafter, the transfected cells are identified and isolated from bulk. They are multiplied and introduced into blastocyst through microinjection. The blastocyst is transferred into uterus of a surrogate mother for further developmental stages. The animals, which are born, are checked for the presence of transgene. The transgenic animal is crossed with a normal one to study the inheritance of introduced foreign gene.

6

Breeding Techniques of Farm Animals

Objectives in the breeding of farm animals, role of animal breeder in improving livestock economy. Existing livestock genetic resources with particular emphasis on production potentials; comparison with foreign breeds and basis for the superior/inferior performance. Breeding policies for cattle, buffaloes, sheep and goats and other livestock species in Pakistan. Strategies for improved use and conservation of livestock. Inbreeding and crossbreeding for livestock improvement: Development and uses of inbred lines in poultry. Heterosis: genetic aspects, exploitation, crossbreeding systems. Detection and elimination of genetic defects and lethals from the stock. Maintenance of performance records, their standardization and optimal use. Large scale genetic evaluation. Interpreting genetic evaluation information.

Practicals

Exercises on the evaluation of productive traits of various breeds of livestock. Exercises on standardization of records, genetic evaluation and selection, Computer usage for animal evaluation. Review paper required. Visits to different breeding farms.

Advanced Animal Breeding

Current animal breeding, estimation of genetic parameters, selection for economic merit, testing and selection within stocks. Role of A.I. and other related techniques in enhancing the genetic progress. Multi Trait Across Country Evaluation (MACE) and its application. Genetic technologies: role of biotechnology in animal production, recombinant DNA techniques, recombinant DNA for the improvement of domestic animals; Micro-manipulation: cloning, homozygous diploids, androgenesis and gynogenesis, sex selection, chimaeras, transfer of genes. The direct detection of genotype, DNA polymorphism. Expected genetic and economic gains. Detecting major genes, basic concepts of marker based analysis, influence of markers on production and reproduction in farm animals, mapping and characterizing Quantitative Trait Loci (QTL), genetic maps. Review of the recent literature in animal breeding.

Statistical Methods in Animal Production

Basic biometrical concepts and definitions. Data description for single and multiple variables. Probability and its distributions. Estimation and hypothesis testing. Concepts and application of regression and correlation, least squares analysis. Introduction to analysis of variance and assumption underlying it. Principles of planning of animal experiments.. Conceptsbehind using various experimental designs such as completely randomized design, randomized complete block designs and Latin square design with and without factorial arrangements. Changeover, crossover and other designs. Relative efficiencies of these designs. Multiple comparisons. Fixed, random and mixed models. Estimation of variance components under various statistical models. Categorical animal data analysis, animal production data management and report preparation.

Introduction to various computer packages for animal data handling and processing.

Numerical exercises on various experimental methods listed in the theory manually and by using computers.

Efficiency of Reproduction in Farm Animals

Genetic basis of variation in reproductive functions. Relationship of reproduction to genetic improvement in domestic animals; Pre-and post natal selection of genetic and reproductive material; the problem of reduced fertility and reproductive efficiency and its relation to lifetime productive efficiency; Measures of reproductive efficiency in farm animals; Genetic and environmental basis of the differences in reproductive efficiency; Fertility level of males for use in artificial insemination; Reduction in estrous period, quiescent heat and anestrus; Embryonic losses; Issues related to prolificacy; Manipulation of reproductive cycle, Feeding and management practices in improving fertility traits; Augmenting reproductive rate, Application of biotechnology for improving reproductive efficiency.

Exploitation of genetic variation in reproductive performance for improvement.

Practice on the slaughter house material and experimental live animals. Rectal palpation of genital organs at various stages of reproductive cycle. Practice in various methods of heat detection; correction of reproductive problems. Hormonal application and other techniques. Practice in pregnancy diagnosis in various livestock species. Measurement of reproductive fitness in males. Determining reproductive efficiency in males and females.

Biometrical Genetics

Statistical concepts in animal genetics. Review of matrix algebra and linear models, elementary probability and its application to Mendelian inheritance. Metric characters, population mean, average effect, breeding value, dominance deviations, interaction deviations. Sources of genetic variation in single and multilocus traits, sources of environmental variation. Estimation of variance components: parent-offspring regression, sib analysis, cross-classified designs, genotype x environment interaction, maternal effects and other genetic models. Estimation of breeding values.

Theory of Inbreeding

The concept of inbreeding and its measurement. Wright's method of path coefficients and variance co-variance chart, Malecot's concept of relationship and inbreeding. Distribution, mean and variance of gene frequency in small populations. Genotypic frequencies with inbreeding. Regular systems of inbreeding. Inbreeding with two loci. Effects of inbreeding on the variance, effective population size. Dangers of inbreeding, inbreeding depression and heterosis. Practical uses of relationship coefficients. Selection and inbreeding, simulation as applied to inbreeding.

Experimental Techniques in Population Genetics

The lifecycle of Drosophila melanogaster, Genetics of embryonic development in Drosophila and Tribolium, different stages, methods of breeding and recording. Methods of genetic analysis with respect to gene action and properties of genetic material. Analytical procedures by means of markers. Verification of theories of population genetics on the colonies of the above species with special emphasis on quantitative traits like bristle number, larval weight, etc. Review of research work done.

Handling and breeding of Drosophila and/or Tribolium species. Experiments of gene frequency, genotypic ratios, Hardy-Weinberg law, and departures from it in small populations. Demonstration of selection and inbreeding experiments for verification of quantitative theory.

Poultry Genetics and Breeding

Basic genetic concepts. Poultry genetic resources, reproductive biology in relation to genetics and breeding, sex determination, genetics of the plumage, skin, growth, meat production, egg production, egg composition, fertility, diseases and behaviour. Genetic controls in selection, genotype by environment interaction, heterosis and inbreeding. Development of commercial synthetic layers and broilers, Molecular and immunogenetic aspects of poultry genetics, genetic engineering, mutations and major variants in the fowl. Breeding for fancy poultry, Breeding for rural poultry production systems. Conservation of poultry genetic resources.

CURRENT USE, RESEARCH AND IMPENDING DEVELOPMENT OF FOODS PRODUCED THROUGH MODERN BIOTECHNOLOGY

Foods produced through modern biotechnology can be categorized as follows:

1. Foods consisting of or containing living/viable organisms, e.g. maize.
2. Foods derived from or containing ingredients derived from GMOs, e.g. flour, food protein products, or oil from GM soybeans.
3. Foods containing single ingredients or additives produced by GM microorganisms (GMMs), e.g. colours, vitamins and essential amino acids.
4. Foods containing ingredients processed by enzymes produced through GMMs, e.g. high-fructose corn syrup produced from starch, using the enzyme glucose isomerase (product of a GMM).

This study, however, makes no attempt to discriminate between the various categories, and the discussion that follows describes the current and future applications of modern biotechnology in the production of crops, livestock, fish and microorganisms in food production.

Crop Breeding and the Introduction of GM crops for food Production

Conventional breeding, especially of crops, livestock and fish, focuses principally on increased productivity, increased resistance to diseases and pests, and enhanced quality with respect to nutrition and food processing. Advances in cellular genetics and cell biology methods in the 1960s contributed to the so-called 'green revolution' that significantly increased varieties of staple food crops containing traits for higher yield and resistance to diseases and pests in a number of both developed and developing countries (Borlaug 2000). A key driver of the green revolution was to improve the potential to provide sufficient food for all. The intensification and expansion of agriculture brought about by these methods and agricultural systems have, however, also resulted in new forms of health and environmental risks through, for example, increased use of agrochemicals and intensified cultivation resulting in soil erosion.

The development of molecular biology in the 1970s and 1980s introduced more direct methods for the analysis of genetic sequences and allowed the identification of genetic markers for desired traits. Such marker-assisted breeding methods are the basis of some current conventional breeding strategies. Whereas modern methods of breeding have significantly increased crop yields over the past 50 years, the future potential of these methods is constrained by the limitations in the natural diversity of trait genotype within crop species and sexual-compatibility boundaries between crop types.

To overcome these problems, a number of interested groups (scientists, farmers, governments, agricultural companies) have since the 1980s considered other means to achieve the objectives of improved yields, sustainable agricultural systems, and improvements in human and animal health and the environment. This includes the use of more modern methods to introduce novel traits, such as tolerance to drought, salt, or pests. To achieve these objectives, various public and, more recently, private research programmes have aimed to improve the understanding of and links between crop performance and molecular genetics. With the development and use of recombinant DNA in the 1980s, a tool to overcome the limitation of species incompatibility was found. Modern biotechnology employs molecular techniques to identify, select and modify DNA sequences for a specific genetic trait (e.g. insect resistance) from a donor organism (microorganism, plant or animal), and transfer the sequence to the recipient organism so that it expresses this trait.

Various transformation methods are used to transfer recombinant DNA into recipient species to produce a GMO. For plants, these include transformation mediated by *Agrobacterium tumefaciens* (a common soil bacterium that contains genetic elements for infection of plants) and biolistics — shooting recombinant DNA placed on microparticles into recipient cells. The methods used in the transformation of various animal species include microinjection, electroporation and germ-line cells (FAO/WHO 2003a). The success rate of transformations in animals tends to be lower than in plants, and to vary from species to species, thus requiring the use of many animals.

Genetic modification is often faster than conventional breeding techniques, as stable expression of a trait is achieved using far fewer breeding generations. It also allows a more precise alteration of an organism than conventional methods of breeding, as it enables the selection and transfer of a specific gene of interest. However, with the present technology, in many cases it leads to random insertion in the host genome, and consequently may have unintended developmental or physiological effects. However, such effects can also occur in conventional breeding and the selection process used in modern biotechnology aims to eliminate such unintended effects to establish a stable and beneficial trait.

It should be noted that conventional breeding programmes directed by the molecular analysis of genetic markers are also of critical importance to modern plant and animal breeding. However, human and environmental health consequences of these techniques are not considered here.

GM Crops Currently in Commercial Production

At present, only a few GM crops are permitted for food use and traded on the international food and feed markets. These include herbicide-and insect-resistant maize (*Bt*[1] maize), herbicide-resistant soybean, rape (canola) oilseed, and insect-and herbicide-resistant cotton (primarily a fibre crop, though refined cottonseed oil is used as food). In addition, several government authorities have approved varieties of papaya, potato, rice, squash, sugar beet and tomato for food use and environmental release. The latter crops, however, are currently grown and traded only in a limited number of countries, mainly for domestic consumption.

The regulatory status of GM crops varies among the countries that permit their use and updates can be found on various web-sites, including those of the Organisation for Economic Co-operation and Development (OECD) and the International Centre for Genetic Engineering and Biotechnology (ICGEB).

Future Trends in GM Crops

The commercial introduction of transgenic crop plants with agronomic traits is often referred to as the first generation of transgenic plants. Further development of GM crops with agronomic traits is continuing, and production of a range of GM crops with enhanced nutritional profiles is also under way (PIFB 2001). Various novel traits are currently being tested in laboratories and field tests in a number of countries. Many of these second-generation GM crops are still in the development stage and are unlikely to enter the market for several years. The key areas of research and development (R&D) in plants are (i) agronomic traits and (ii) altered nutrition and composition.

PPAgronomic Traits

Pest and disease resistance: In the short term, most newly commercialized GM crops will continue to concentrate on agronomic traits, especially herbicide resistance and insect resistance and, indirectly, yield potential (PIFB 2001). R&D in this area aims to:

- introduce herbicide-resistance traits in a broader range of varieties of maize, soybean and canola;
- broaden the range of herbicides that can be used in combination with the transgenic herbicide-resistant crop, such as introduction of tolerance to the herbicides bromoxynil, oxynil and sulfonylurea; and

- stack novel genes for insect resistance in plants, such as novel *Bt* variants containing different toxins.

Virus resistance: Virus resistance could be extremely important to improving agricultural productivity (Thompson 2003). Field tests of the following virus-resistant crops are currently being conducted in various parts of the world: sweet potato (feathery mottle virus); maize (maize streak virus); and African cassava (mosaic virus). These crops may be available for commercialization within the next 3–5 years. Because of its complex genome, work on wheat resistant to the barley yellow-dwarf virus has made little progress and is still undergoing laboratory investigation. Resistance to nematodes (root worms) in a GM potato has also been achieved.

Altered Nutrition and Composition

Vitamin-A-enhanced rice: The best-known example of a GM crop conferring enhanced nutritional properties is rice containing a high level of beta-carotene — a vitamin A precursor (so-called 'golden rice') (Potrykus 2000). Vitamin A is essential for increasing resistance to disease, protecting against visual impairment and blindness, and improving the chances of growth and development. Vitamin A deficiency (WHO/UNICEF 1995) is a public health problem that contributes to severe illness and childhood mortality. This preventable condition increases the burden of disease on the health systems of developing countries. A number of strategies have been suggested for combating vitamin A deficiency, including dietary approaches (e.g. fortification of foods) and supplementation via pills (WHO 2000c). Within the context of improving the supply of vitamin A, the usefulness of vitamin A-enhanced rice has been discussed in various forums, such as an electronic forum coordinated by the Food and Agriculture Organization of the United Nations (FAO) in 2000 (FAO 2000).

Vitamin A-enhanced rice and maize varieties are at present being developed for cultivation in developing countries. Current efforts are aimed at ensuring that vitamin A in rice can be absorbed efficiently in the human gut. Once this is resolved, 300 grams of transgenic rice could make a significant contribution to the daily human requirement for vitamin A.

'High iron' rice : Prevalence of iron deficiency is very high in those parts of the world in which rice is the daily food staple (WHO 2000a). This is because rice has a very low iron content. Transgenic rice seeds with the iron-carrier protein ferritin from soy were found to contain twice as much iron as seeds of non-transformed rice (Gura 1999). Rice has been transformed with three genes which increase iron storage in rice kernels and iron absorption from the digestive tract (Lucca et al. 2002).

Improved protein content : Researchers are also investigating methods that could improve the protein content of staple vegetables, such as cassava, plantain and potato (PIFB 2001). Results from greenhouse trials show that these tubers have 35–45% more protein, and enhanced levels of essential amino acids.

Removing allergens and antinutrients: Cassava roots naturally contain high levels of cyanide. As they are a staple food in tropical Africa, this has led to high blood-cyanide levels which have harmful effects. Application of modern biotechnology to decrease the levels of this toxic chemical in cassava would reduce its preparation time. In potatoes, insertion of an invertase gene from yeast reduces the natural levels of glycoalkaloid toxin (Buchanan et al. 1997).

The allergenic protein in rice has been reduced by modifying its biosynthetic pathway (PIFB 2001). The significance in human allergenicity of these lower levels has not been demonstrated. There is also work to reduce allergenicity in wheat (Buchanan et al. 1997). This work involves inserting a thioredoxin-biosynthesis gene to break the disulfide bonds in the offending protein but without interfering with the functionality of the wheat proteins.

Altered starch and fatty acid profile: In the quest to provide healthier foods, there is an effort to increase the starch content of potatoes so that they absorb less fat during frying (PIFB 2001). To create healthier fats, the fatty-acid composition of soy and canola has been altered to produce oils with reduced levels of saturated fats. R&D is currently focusing on GM soybean, oilseed rape and oil palm (PIFB 2001). Two GM crops of this nature have been approved in the United States of America (USA) for growing and food/feed use – high oleic acid soy and high lauric acid oilseed rape (Agbios 2005). High oleic acid soy is also permitted as food in Australia and Canada. R&D is in the early stages with respect to oils with improved nutritional value.

Increased antioxidant content: The lycopene and lutein contents of tomatoes have been increased as have isoflavones in soy (WHO 2000c). These phytonutrients are known to improve health or prevent disease. Research in this area is at a relatively early stage of development, as knowledge of phytonutrients is limited and not all phytonutrients are beneficial.

Environmental stresses: Tolerance to environmental-stress factors through genetic modification is an area that is in the early stages of R&D (PIFB 2001). Resistance to salinity and drought are being researched intensively. Salinity is estimated to affect 20% of agricultural land and 40% of irrigated land worldwide. Salt and drought tolerance involve numerous genes interacting in a complex manner. Owing to this multigenic character, conventional breeding techniques have had little success in the generation of salt-or drought-tolerant varieties. Salt tolerance may be conferred to sensitive crops by the transfer of multiple genes linked to a relevant pathway from a tolerant crop. The likely time frame for commercialization for such GM crops is unknown.

Tolerance to aluminium (a growth-limiting factor in acid soils) is in the early phase of R&D for several crops, including papaya, tobacco, rice and maize, but they are not expected to be in commercial use for several years.

Attempts have been made to improve the photosynthetic system in plants through genetic modification. Crops such as maize and sugar cane are more efficient in converting energy into sugars than most broadleaf crop plants. By introducing genes for more efficient photosynthesis from one crop to another, efficiency could be improved by 10% with an enhancement in yield. The likely time frame of commercialization is unknown.

Male-sterility traits have been introduced for obtaining 100% hybrid sowing-seed for the purposes of environmental containment of GM crops. Various male-sterile maize varieties have been approved for market introduction in the USA. In addition, various male-sterile rapeseed and canola varieties have been approved for environmental release and food use in the European Union (EU), Canada and the USA. Another strategy for containing gene flow between plants attempts to introduce asexual seed propagation in crops (seed production without the need to pollinate). None of the above-mentioned strategies has proved applicable to all crop species, and a combination of approaches may prove most effective.

Livestock and Fish

In terms of food production, the application of modern biotechnology to livestock falls into two main areas: animal production and human nutrition. Many of the applications discussed below are in the early stages of R&D.

Fish

The projected increasing demand for fish suggests that GM fish may become important in both developed and developing countries. Enhanced-growth Atlantic salmon containing a growth hormone gene from Chinook salmon is likely to be the first GM animal on the food market (FAO/WHO 2003a). These fish grow 3–5 times faster than their non-transgenic counterparts, to reduce production time and increase food availability. At least eight other farmed fish species have been genetically modified for growth enhancement. Other fish in which genes for growth hormones have been experimentally introduced include grass carp, rainbow trout, tilapia and catfish (PIFB 2003; PIFB/FDA 2003). In all cases, the growth-hormone genes are of fish origin.

To address some of the practical problems of aquaculture, research attempts are seeking to improve disease resistance by producing Atlantic salmon with a rainbow trout lysozyme cDNA. Lysozyme has antimicrobial properties against fish pathogens such as *Vibrio, Aeromonas* and *Yersinia*. Another type of antimicrobial protein (silk moth cecropin) is under investigation in catfish (Dunham et al. 2002). This would improve catfish resistance to diseases such as enteric septicaemia.

The farming of carnivorous fish species, such as trout and salmon, has led to overfishing of sand eels and capelin. To tackle this problem, research is looking into the possibility of altering the metabolism of these species by improving their digestion of carbohydrates, to enable a shift to a more plant-based diet.

Lack of cold tolerance in warm-water species such as the common carp and tilapia can lead to significant stock losses in winter. The suggestion of work in this area is to alter the molecular conformation of lipids, thus increasing membrane fluidity.

To extend the geographical range of fish farming, an antifreeze gene from one fish species is transferred to the species of interest. Although freeze-resistant strains of Atlantic salmon have been produced, the level of antifreeze protein secreted by the salmon was insufficient to have a significant impact on the freezing point of blood (Fletcher et al. 2002).

Livestock and Poultry

Foods derived from GM livestock and poultry are far from commercial use. Several growth-enhancing novel genes have been introduced into pigs that have also affected the quality of the meat, i.e. the meat is more lean and tender (FAO/WHO 2003a). This research was initiated over a decade ago, but owing to some morphological and physiological effects developed by the pigs, these have not been commercialized.

Many modifications to milk have been proposed that either add new proteins to milk or manipulate endogenous proteins (PIFB 2002b). Recently, researchers from New Zealand developed GM cows that produce milk with increased levels of casein protein. Use of such protein-rich milk would increase the efficiency of cheese production. Other work aims to reduce the lactose content of milk, with the intent of making milk available to the population of milk-intolerant individuals.

Other applications of genetic modification in animal production in the early stages of R&D include improvement of disease resistance, increased birth rates in sheep, altered sex ratio in poultry, increased egg production in poultry by creating two active ovaries, and improved feed conversion in the 'enviropig' (environmentally friendly pigs that excrete less phosphorus). Most of this work is still theoretical and therefore estimates of time frames for possible commercial introductions of any of these applications are unavailable.

FACTORY FARMING

Factory farming is a term referring to the process of raising livestock in confinement at high stocking density, where a farm operates as a factory — a practice typical in industrial farming by agribusinesses. The main product of this industry is meat, milk and eggs for human consumption. However, there have been issues regarding whether factory farming is sustainable and ethical.

Confinement at high stocking density is one part of a systematic effort to produce the highest output at the lowest cost by relying on economies of scale, modern machinery, biotechnology, and global trade. Confinement at high stocking density requires antibiotics and pesticides to mitigate the spread of disease and pestilence exacerbated by these crowded living conditions. In addition, antibiotics are used to stimulate livestock growth by killing intestinal bacteria. There are differences in the way factory farming techniques are practiced around the world. There is a continuing debate over the benefits and risks of factory farming. The issues include the efficiency of food production; animal welfare; whether it is essential for feeding the growing global human population; the environmental impact and the health risks.

Agriculture adopted more intensive methods during the eighteenth century. With this growth in production best characterized by the Agricultural Revolution, where improvements in farming techniques allowed for significantly improved yields, and supported the urbanization of the population during the Industrial Revolution.

Innovations in agriculture beginning in the late nineteenth century paralleled developments in mass production in other industries. The identification of nitrogen and phosphorus as critical factors in plant growth led to the manufacture of synthetic fertilizers, making possible more intensive types of agriculture.

The first animals to be factory farmed were chickens. The discovery of vitamins and their role in animal nutrition, in the first two decades of the twentieth century, led to vitamin supplements, which allowed chickens to be raised indoors. The discovery of antibiotics and vaccines facilitated raising livestock in larger numbers by reducing disease. Chemicals developed for use in World War II gave rise to synthetic pesticides. Developments in shipping networks and technology have made long-distance distribution of agricultural produce feasible.

According to the BBC, factory farming in Britain began in 1947 when a new Agriculture Act granted subsidies to farmers to encourage greater output by introducing new technology, in order to reduce Britain's reliance on imported meat. The United Nations writes that intensification of animal production was seen as a way of providing food security. In 1960s North America, pigs and cows began to be raised on factory farms. This practice then spread to Western Europe. In Britain, the agriculture correspondent of *The Guardian* wrote in 1964:

Factory farming is greatly debated throughout Australia, with several people disagreeing with the methods and ways in which the animals in factory farms are treated. Often animals are under stress from being kept in confined spaces, so will attack each other. This results in them having beaks, tails and teeth removed. Many piglets will die of shock after having their teeth and tails removed. This is due to the fact that painkilling medicines are not used in these operations. Others say that factory farms are a great way to gain space, with animals such as chickens being kept in spaces smaller than an A4 page. From its American and West European heartland factory farming became globalised in the later years of the twentieth century and is still expanding and replacing traditional practices of stock rearing in an increasing number of countries. In 1990 factory farming accounted for 30% of world meat production. By 2005 this had risen to 40%.

Nature of the Practice

Factory farms hold large numbers of animals, typically cows, pigs, turkeys, or chickens, often indoors, typically at high densities. The aim of the operation is to produce as much meat, eggs, or milk at the lowest possible cost. Food is supplied in place, and a wide variety of artificial methods are employed to maintain animal health and improve production, such as the use of antimicrobial agents, vitamin supplements, and growth hormones. Physical restraints are used to control movement or actions regarded as undesirable. Breeding programs are used to produce animals more suited to the confined conditions and able to provide a consistent food product.

The distinctive characteristic of factory farms is the intense concentration of livestock. At one farm (Farm 2105) run by Carrolls Foods of North Carolina, the second-largest pig producer in the U.S., twenty pigs are kept per pen and each confinement building or "hog parlor" holds 25 pens. The company's chief executive officer, F.J. "Sonny" Faison, has said: "It's all a supply-and-demand price question ... The meat business in this country is just about perfect, uncontrolled supply-and-demand free enterprise. And it continues to get more and more sophisticated, based on science. Only the least-cost producer survives in agriculture." Faison states:

The large concentration of animals, animal waste, and the potential for dead animals in a small space poses ethical issues. It is recognized that some techniques used to sustain intensive agriculture can be cruel to animals. As awareness of the problems of intensive techniques has grown, there have been some efforts by governments and industry to remove inappropriate techniques. Farm Sanctuary defines factory farming as "an attitude that regards animals and the natural world merely as commodities to be exploited for profit." The group contends "this attitude has led to institutionalized animal cruelty, massive environmental destruction and resource depletion, and animal and human health risks."

In the UK, the Farm Animal Welfare Council was set up by the government to act as an independent advisor on animal welfare in 1979 and expresses its policy as five freedoms: from hunger & thirst; from discomfort; from pain, injury or disease; to express normal behaviour; from fear and distress.

There are differences around the world as to which practices are accepted and there continue to be changes in regulations with animal welfare being a strong driver for increased regulation. For example, the EU is bringing in further regulation to set maximum stocking densities for meat chickens by 2010, where the UK Animal Welfare Minister commented, "The welfare of meat chickens is a major concern to people throughout the European Union. This agreement sends a strong message to the rest of the world that we care about animal welfare."

However, given the assumption that intensive farming techniques are a necessity, it is recognized that some apparently cruel techniques are better than the alternative. For example, in the UK, de-beaking of chickens is deprecated, but it is recognized that it is a method of last resort, seen as better than allowing vicious fighting and ultimately cannibalism. Between 60 and 70 percent of six million breeding sows in the U.S. are confined during pregnancy, and for most of their adult lives, in 2 ft (0.61 m) by 7 ft (2.1 m) gestation crates. According to pork producers and many veterinarians, sows will fight if housed in pens. The largest pork producer in the U.S. said in January 2007 that it will phase out gestation crates by 2017. They are being phased out in the European Union, with a ban effective in 2013 after the fourth week of pregnancy. With the evolution of factory farming, there has been a growing awareness of the issues amongst the wider public, not least due to the efforts of animal rights and welfare campaigners. As a result gestation crates, one of the more contentious practices, are the subject of laws in the U.S., Europe and around the world to phase out their use as a result of pressure to adopt less confined practices'.

Human Health Impact

According to the U.S. Centres for Disease Control and Prevention (CDC), farms on which animals are intensively reared can cause adverse health reactions in farm workers. Workers may develop acute and chronic lung disease, musculoskeletal injuries, and may catch infections that transmit from animals to human beings (such as tuberculosis). Pesticides are used to control organisms which are considered harmful and they save farmers money by preventing product losses to pests.

In the US, about a quarter of pesticides used are used in houses, yards, parks, golf courses, and swimming pools and about 70% are used in agriculture. However, pesticides can make their way into consumers' bodies which can cause health problems. One source of this is bioaccumulation in animals raised on factory farms.

The CDC writes that chemical, bacterial, and viral compounds from animal waste may travel in the soil and water. Residents near such farms report problems such as unpleasant smell, flies and adverse health effects.

The CDC has identified a number of pollutants associated with the discharge of animal waste into rivers and lakes, and into the air. The use of antibiotics may create antibiotic-resistant pathogens; parasites, bacteria, and viruses may be spread; ammonia, nitrogen, and phosphorus can reduce oxygen in surface waters and contaminate drinking water; pesticides and hormones may cause hormone-related changes in fish; animal feed and feathers may stunt the growth of desirable plants in surface waters and provide nutrients to disease-causing micro-organisms; trace elements such as arsenic and copper, which are harmful to human health, may contaminate surface waters.

In the European Union, growth hormones are banned on the basis that there is no way of determining a safe level. The UK has stated that in the event of the EU raising the ban at some future date, to comply with a precautionary approach, it would only consider the introduction of specific hormones, proven on a case by case basis. In 1998, the European Union banned feeding animals antibiotics that were found to be valuable for human health. Furthermore, in 2006 the European Union banned all drugs for livestock that were used for growth promotion purposes. As a result of these bans, the levels of antibiotic resistance in animal products and within the human population showed a decrease.

The various techniques of factory farming have been associated with a number of European incidents where public health has been threatened or large numbers of animals have had to be slaughtered to deal with disease. Where disease breaks out, it may spread more quickly, not only due to the concentrations of animals, but because modern approaches tend to distribute animals more widely. The international trade in animal products increases the risk of global transmission of virulent diseases such as swine fever, BSE, foot and mouth and bird flu.

Methicillin-resistant Staphylococcus aureus (MRSA) has been identified in pigs and humans raising concerns about the role of pigs as reservoirs of MRSA for human infection. One study found that 20% of pig farmers in the United

States and Canada in 2007 harbored MRSA. A second study revealed that 81% of Dutch pig farms had pigs with MRSA and 39% of animals at slaughter carried the bug were all of the infections were resistant to tetracycline and many were resistant to other antimicrobials. A more recent study found that MRSA ST398 isolates were less susceptible to tiamulin, an antimicrobial used in agriculture, than other MRSA or methicillin susceptible *S. aureus*. Cases of MRSA have increased in livestock animals. CC398 is a new clone of MRSA that has emerged in animals and is found in intensively reared production animals (primarily pigs, but also cattle and poultry), where it can be transmitted to humans. Although being dangerous to humans CC398 is often asymptomatic in food-producing animals.

A 2011 study reported that according to a nationwide study nearly half of the meat and poultry sold in U.S. grocery stores—47 percent—was contaminated with S. aureus, and more than half of those bacteria—52 percent—were resistant to at least three classes of antibiotics. Although Staph should be killed with proper cooking, it may still pose a risk to consumers through improper food handling and cross-contamination in the kitchen. The senior author of the study said, "The fact that drug-resistant S. aureus was so prevalent, and likely came from the food animals themselves, is troubling, and demands attention to how antibiotics are used in food-animal production today." In April 2009, lawmakers in the Mexican state of Veracruz accused large-scale hog and poultry operations of being breeding grounds of a pandemic swine flu, although they did not present scientific evidence to support their claim. A swine flu which quickly killed more than 100 infected persons in that area, appears to have begun in the vicinity of a Smithfield subsidiary pig CAFO (concentrated animal feeding operation).

Animal Health Impact

Confinement and overcrowding of animals results in a lack of exercise and natural locomotory behaviour, which weakens their bones and muscles. An intensive poultry farm provides the optimum conditions for viral mutation and transmission – thousands of birds crowded together in a closed, warm, and dusty environment is highly conducive to the transmission of a contagious disease. Selecting generations of birds for their faster growth rates and higher meat yields has left birds' immune systems less able to cope with infections and there is a high degree of genetic uniformity in the population, making the spread of disease more likely. Further intensification of the industry has been suggested by some as the solution to avian flu, on the rationale that keeping birds indoors will prevent contamination. However, this relies on perfect, fail-safe biosecurity – and such measures are near impossible to implement. Movement between farms by people, materials, and vehicles poses a threat and breaches in biosecurity are possible. Intensive farming may be creating highly virulent avian flu strains. With the frequent flow of goods within and between countries, the potential for disease spread is high. Confinement and overcrowding of animals' environment presents the risk of contamination of the meat from viruses and bacteria. Feedlot animals reside in crowded conditions and often spend their time standing in their own waste. A dairy farm with 2,500 cows may produce as much waste as a city of 411,000 people, and unlike a city in which human waste ends up at a sewage treatment plant, livestock waste is not treated. As a result, feedlot animals have the potential of exposure to various viruses and bacteria via the manure and urine in their environment. Furthermore, the animals often have residual manure on their bodies when they go to slaughter.

Confinement at high stocking density requires antibiotics and pesticides to mitigate the spread of disease and pestilence exacerbated by these crowded living conditions. In addition, antibiotics are used to stimulate livestock growth by killing intestinal bacteria. According to a February 2011 FDA report, nearly 29 million pounds of antimicrobials were sold in 2009 for both therapeutic and non-therapeutic use for all farm animal species. The Union of Concerned Scientists estimates that 70% of that amount is for non-therapeutic use

Environmental Impact

Concentrating large numbers of animals in factory farms is a major contribution to global environmental degradation, through the need to grow feed (often by intensive methods using excessive fertiliser and pesticides), pollution of water, soil and air by agrochemicals and manure waste, and use of limited resources (water, energy).

Livestock production is also particularly water-intensive in indoor, intensive systems. Eight percent of global human water use goes towards animal production, including water used to irrigate feed crops.

Industrial production of pigs and poultry is an important source of GHG emissions and is predicted to become more so. On intensive pig farms, the animals are generally kept on concrete with slats or grates for the manure to drain through. The manure is usually stored in slurry form (slurry is a liquid mixture of urine and faeces). During storage on farm, slurry emits methane and when manure is spread on fields it emits nitrous oxide and causes nitrogen pollution of land and water. Poultry manure from factory farms emits high levels of nitrous oxide and ammonia.

Organic pig meat production has a lower global warming potential per kg than does intensive pig meat production. The energy input for free-range poultry meat and eggs is higher than for factory-farmed poultry meat and eggs, but GHG emissions are lower.

Environmental impacts of factory farming can include:

- Deforestation for animal feed production
- Unsustainable pressure on land for production of high-protein/high-energy animal feed
- Pesticide, herbicide and fertilizer manufacture and use for feed production
- Unsustainable use of water for feed-crops, including groundwater extraction
- Pollution of soil, water and air by nitrogen and phosphorus from fertiliser used for feed-crops and from manure
- Land degradation (reduced fertility, soil compaction, increased salinity, desertification)
- Loss of biodiversity due to eutrophication, acidification, pesticides and herbicides
- Worldwide reduction of genetic diversity of livestock and loss of traditional breeds
- Species extinctions due to livestock-related habitat destruction (especially feed-cropping).

Animal Welfare Impact

Animal welfare impacts of factory farming can include:

- Close confinement systems (cages, crates) or lifetime confinement in indoor sheds
- Discomfort and injuries caused by inappropriate flooring and housing
- Restriction or prevention of normal exercise and most of natural foraging or exploratory behaviour
- Restriction or prevention of natural maternal nesting behaviour
- Lack of daylight or fresh air and poor air quality in animal sheds
- Social stress and injuries caused by overcrowding
- Health problems caused by extreme selective breeding and management for fast growth and high productivity
- Reduced lifetime (longevity) of breeding animals (dairy cows, breeding sows)
- Fast-spreading infections encouraged by crowding and stress in intensive conditions
- Debeaking (beak trimming or shortening) in the poultry and egg insustry to avoid pecking in overcrowded quarters
- Forced and over feeding (by inserting tubes into the throats of ducks) in the production of foie gras.

BREEDING BEEF CATTLE

There are many things to consider when raising cattle, particularly beef cattle, and one of these is actually breeding beef cattle. Many cattle farmers spend a certain amount of their time trying to breed their cattle not only to be able to increase the herd but to also produce high quality, healthy, and close-to-perfect calves.

They know that consumers and other cattle breeders only want the best in their herd, so breeding beef cattle should be a major addition to one's agenda when engaged in cattle farming. Unfortunately, some farmers tend to forget the most basic tips and techniques to breeding high quality cattle and have failed to renew the expenses made for ill-fitted cows and calves with health problems at birth.To be able to avoid this, here are some helpful tips on how to effectively breed beef cattle:

Breeding beef cattle means knowing how to balance between genetics and herd management. With this in mind, remember to be selective with one's own herd by learning how to distinguish the good animals from those with health and bone structure problems. Keeping the best and raising the best is the key to successful breeding because it not only produces cows with the best meat but also reduces the time, effort, and money spent on work at breeding time.

Bulls and heifers should be examined once in a while, especially when the cows are starting to show pregnancy signs and are ready to conceive. Try certain techniques to increase your chances of producing well-rounded cattle. For example, make sure to choose heifers with large pelvises and a bull whose mother also had a large pelvis.

There are two ways to go for when breeding beef cattle depending on what your preferences are. You can either artificially inseminate the cows or keep a good bull to breed with your cows.

If you've chosen the second method, have your bull/s examined for STDs, health problems, and once in a while for general physical check-ups, sperm count, and scrotum circumference to be sure that your cows will easily conceive.

Choosing the bulls for breeding also means choosing those that have been known to produce well-rounded calves. You can go around, ask other cattle farmers, and see if they have good bulls up for sale that you can keep in your herd.

Top Five Ways for Breeding Finches

Anyone that has finches for pets will eventually consider breeding their birds. The delight you get from your pets makes you want to see your birds have little ones, and watch them grow.

To begin breeding finches you need to get your birds a nest. Some people provide store-bought nests for their pets, some people construct the nests for them, and some people will provide the animals with the materials to build their own nests. Do not be upset if your birds do not like the first nest you give them, this is absolutely normal. Just remove the one they did not like and try another one. Continue to replace them until the birds accept one, and after that you will always know which nest to provide for your breeding finch.

Zebra finch breeders often provide their birds with wicker basket nests that are full of material to encourage their breeders. When these animals breed in the wild they will look for a place where they can hide completely, and this makes them feel safe. If you give them the ability to feel safe this will encourage breeding.

If your pets are not breeding you can try buying a heater from the pet supply, like they place in reptile cages; the heater will make the animals feel better and be healthier. Healthy birds are more likely to breed than sick ones.

You can try providing your animals with plenty of material to put in their nest. Sisal fiber is an excellent choice in materials as the finches breeders who do this on a regular basis will tell you.

One of the best things you can do for your pets is to make their environment as close to the natural environment they would have been in. They need to feel safe and secure, and they need to have what they consider a good food supply. This is not hard to accomplish, you can keep the feeder full for the animals, and plants make them feel like they have places to hide from danger. Buy fake plants for this purpose, or you can try providing them with cover by placing a blanket on the sides of their cage.

If you have children in your home, do not let them bump the cage a lot, or poke their fingers or other objects through the bars. If the birds are frequently disrupted by their cage being jarred around, then they will not feel safe and secure.

When you notice the male fussing with the nest, and then doing their mating song and dance, you should notice an egg in the nest in the next couple of days. The female may not begin to sit on the egg until she has laid 8 eggs. This is called a clutch of eggs, and she can only lay one per day. After they begin to sit on the eggs you will have hatching in 10 to 18 days. The parents will give the young all the care they need.

Learn the joy of Finch companionship and how to buy, keep and raise healthy Finches. Come find all the information you need at Zebra Finch Breeders. I am Ralph Siskin and I have been raising and learning about Finches for quite some time. I love these birds and want to share what I know to other like minded Finch lovers or people just interested in birds.

Breeding Dogs-The Revolutionary New Methods

Dog breeding is a controversial subject. Many people have heard terrible stories of female dogs being bred time after time for profit and then being abandoned by their owners when they are no longer able to have puppies. Bearing this in mind, it is easy to see why some people may view dog breeding as a negative thing. However, many breeders have a genuine love of the breed(s) they work with and this is what motivates them to get into dog breeding in the first place.

So, if you are considering breeding dogs, the questions you need to ask yourself are-What is motivating you? Do you wish to become a professional breeder, or just mate the family pet? It is not enough to say you simply wish to make some money or you think puppies are cute-puppies are cute, but they are also extremely hard work and cost an absolute fortune to rear.

Consider the cost of finding a suitable stud for your bitch, then all the costs associated with pregnancy: vet bills, adequate food, etc. This will be expensive enough, but any such costs will increase ten-fold once the pups are born (bearing in mind that some breeds can have up to 12 babies). Moreover, if a bitch needs a caesarean to deliver her brood, the fee can be astronomical and you need to be prepared for this.

For professional breeders, mating dogs is a full-time job. They have an incredible commitment to the job and an inherent love of dogs, and this is why they go into the job. Obviously, money is a motivating factor for such people-they have to make a living-but it is not THE motivating factor.

A further thing you need to consider if you are to breed your bitch is time. Not only does a pregnant dog cost money, she requires adequate exercise and will need several short walks every day. Also, in the days leading up to the birth (known as 'the whelping period'), you will need to be constantly present for when labour starts. The newborn pups will also need close attention once they arrive, and so it is imperative that you, or someone else in the family, has the time to devote to all of this.

A responsible dog breeder always makes sure that their dog(s) receive all the relevant innoculations and the pups also. If there are 10-12 pups, this can cost a fortune, but it is important that it is done in order to protect the puppies' health. Another consideration in the mess that pups create-initially, you will need a box big enough to house them, and bedding will need to be changed several times every day. When the pups are old enough to walk about, they will not be house trained and will do their business anywhere-therefore, you need to place them in a room with a hard floor that can be easily cleaned. The pups will also need a secure outside space in which they can run, play and get exercise.

Finally, if you do decide to mate your bitch, do your research on the particular breed-ignorance may put your dog and their babies at risk. Also, when you come to sell the puppies, buyers will ask all sorts of questions and you need to be able to answer them. If you are thinking of becoming a professional breeder, you may need a licence-contact your local government office for advice.

NEW ANIMAL BREEDING TECHNIQUES AND THEIR APPLICATION

The new biotechnologies of gene transfer, in-vitro production, cloning and sexing of embryos have been developed and are being refined with efficiencies suitable for use in animal agriculture. Efficient in-vitro systems for maturing oocytes and capacitating spermatozoa, for fertilizing and developing the embryos have resulted in commercial in-vitro production of embryos. Cloning of embryos by nuclear transfer has been accomplished for sheep, cattle, pigs and rabbits with nuclear material supplied by embryos as late as the 120-cell stage in sheep. Embryos have been recloned but much research is needed to increase the efficiency of this procedure. Research is needed to develop the use of cultured cells in embryo cloning so that the number of clones may be increased to thousands or millions. Embryos of most species can be sexed in a non-damaging way with male specific antibodies and a more efficient method, amplified DNA hybridization, is beginning to be tested commercially.

Transgenic embryos or offspring have been produced for mice, rats, rabbits, chickens, fish, sheep, pigs and cattle. Genes can be targeted for expression in specific tissues but more efficient methods and a better understanding of the genes to be transferred as well as control by man of the time and tissue of specific gene expression are needed. Before many transgenic animals of value can be made, we must know which genes to introduce. Presently there is a poor understanding of the genes influencing animal growth, efficiency of growth, environmental adaptation, meat, milk or egg composition or animal disease resistance. Their identification will come from badly needed efforts to map the genome of domestic animals. These and other new technologies promise to change livestock breeding drastically in the next decade.

Breeding (Animal)

The application of genetic principles to improving heredity for economically important traits in domestic animals. Examples are improvement of milk production in dairy cattle, meatiness in pigs, feed requirements or growth rate in beef cattle, and egg production in chickens. Selection permits the best parents to leave more offspring in the next generation than do poor parents.

Selection is the primary tool for generating directed genetic changes in animals. It may be concentrated on one characteristic, may be directed independently on several traits, or may be conducted on an index or total score which includes information on several traits. In general, the third method is preferable when several important heritable traits need attention. In practice, selection is likely to be a mixture of the second and third methods.

Heritability, the fraction of the total variation in a trait that is due to additive genetic differences, is a key parameter in making decisions in selection. Most traits are strongly to moderately influenced by environmental or managemental differences. Therefore, managing animals to equalize environmental influences on them, or statistically adjusting for environmental differences among animals, is necessary to accurately choose those with the best inheritance for various traits.

The improvement achieved by selection is directly related to the accuracy with which the breeding values of the subjects can be recognized. Accuracy, in turn, depends upon the heritabilities of the traits and upon whether they can be measured directly upon the subjects for selection (mass selection), upon their parents (pedigree selection), upon their brothers and sisters (family selection), or upon their progeny (progeny testing). For traits of medium heritability, the following sources of information are about equally accurate for predicting breeding values of subjects: (1) one record measured on the subject; (2) one record on each ancestor for three previous generations; (3) one record each on five brothers or sisters where there is no environmental correlation between family members; and (4) one record each on five progeny having no environmental correlations, each from a different mate.

Propagation of improved animal stocks is achieved primarily with purebred strains descended from imported or locally developed groups or breeds of animals which have been selected and interbred for a long enough period to be reasonably uniform for certain trademark characteristics, such as coat colour. Because the number of breeding animals is finite and because breeders tend to prefer certain bloodlines and sires, some inbreeding occurs within the pure breeds, but this has not limited productivity in most of these breeds. Crossbreeding makes use of the genetic phenomenon of heterosis. Heterosis is improved performance of crossbred progeny, exceeding that of the average performance of their parents. Most commercial pigs, sheep, and beef cattle are produced by crossbreeding.

Advances in a variety of technologies have application for improvement of domestic animals, including quantitative genetics, reproductive physiology, and molecular genetics. Quantitative geneticists use statistical and genetic information to improve domestic animals. Typically a statistical procedure is used to rank animals based on their estimated breeding values for traits of economic importance. The statistical procedures used allow ranking animals across herds or flocks, provided the animals in different herds or flocks have relatives in common. The primary contribution of reproductive physiology to genetic improvement is to reduce the generation interval. If genetic improvement is increasing at the same rate per generation, more generations can be produced for a fixed time, and thus more gain per unit of time. The most important development was artificial insemination, which allows extensive use of superior males. Another development was embryo transplantation, which allows more extensive use of females. Cloning is a relatively new technique, by which whole and healthy animals have been produced that have the same DNA as the animal from which the cells were taken.

Due to advances in molecular genetics, knowledge is increasing regarding the location of genes on chromosomes and the distance between the genes. In domestic animals, polymorphisms (changes in the order of the four bases) that are discovered in the DNA may be associated with economic traits. When the polymorphisms are associated with or code for economic traits, they are called quantitative trait loci (QTL). When a few or several quantitative trait loci are known that control a portion of the variability in a trait, increasing the frequencies of favourable alleles can enhance the accuracy of selection and augment production. Another use of molecular genetics is to detect the genes that code for genetically predetermined diseases.

An example is the bovine leukocyte deficiency gene, which does not allow white blood cells to migrate out of the blood supply into the tissues to fight infection. The calves perish at a young age. Screening all sires that enter artificial breeding organizations and not using sires that transmit the defect has effectively controlled this condition.

Genetic Engineering

The last four years of the twentieth century witnessed the most rapid adoption of a new technology in history. Since 1996, millions of acres of farmland have been planted with genetically engineered (GE) crops—mainly corn, soybeans, and cotton. This means that genetically engineered organisms (GEOs) are being released to the environment on a massive scale, an event unprecedented in the 3.8 billion year history of life on this planet. This technological upheaval happened virtually without public debate, while our government played the role of enthusiastic promoter, rather than cautious regulator, of this radically new and environmentally hazardous technology. Genetic engineering is a new technology that can combine genes from totally unrelated species in ways not possible using conventional breeding methods. Genes from an animal, say, a fish, can be put into a plant, a strawberry for instance. An attempt to "improve" strawberries by inserting a gene from an Arctic fish has in fact been discussed. The fish gene is supposed to make the strawberries more resistant to frost by causing the strawberry plant to produce a form of antifreeze which the fish normally produce to endure cold ocean conditions.

Over 60 percent of all processed foods purchased by U.S. consumers are manufactured with GE ingredients. Some corn and potatoes have even been genetically engineered to contain a gene from Bt bacteria which causes every cell of the plants to produce an insecticidal toxin. Yet there is no labelling of these or any GE foods as being genetically engineered, because the U.S. Food and Drug Administration (FDA) considers the GEOs from which these foods are made to be "substantially equivalent" to the non-genetically engineered plant from which the GEOs are derived.

The doctrine of substantial equivalence is pure pretext and rationalization with no basis in science. Yet, in a remarkable display of arrogance, the supporters of genetic engineering accuse their critics of not basing their objections on "sound science."

The FDA also uses this "substantial equivalence" rationalization as an excuse to avoid any effective testing of GE foods to determine their safety. Such testing might seriously delay or even prevent the introduction of GE crops into the marketplace.

From the time in the early 1970s when advances in molecular biology led to the development of the techniques we call genetic engineering, until the mid 1990s, the organisms produced by genetic engineering were nearly all confined to laboratories or controlled factory settings. During this time there were almost no releases of genetically engineered organisms to the environment, as genetic engineering was used in basic research and to produce medically useful substances such as insulin. The unrestrained expansion of genetic engineering into agriculture during the past four years changed all that. By 1999 almost 80 million acres of North American farmland were planted with genetically engineered seed. This means massive releases of GEOs to the environment are now taking place. Genetic engineering now poses a very grave threat to the natural environment.

Historic Turning Point

We are now at a turning point in history. We can continue to allow the virtually unrestricted release of genetically engineered organisms to the environment, or we can bring this technology under strict control.

If we continue on our present path of unrestricted releases of GEOs, we will eventually live in a genetically engineered world, as the genome of each species now on earth is either deliberately altered by genetic engineering or indirectly altered by inheritance of transgenes from a genetically engineered organism. In such a world there would be nothing left of living nature, as every species would have been deprived of its genetic integrity, and every ecosystem would thereby have been irreversibly disrupted.

Of special concern to environmentalists should be the fact that trees are now being genetically engineered, and that it is proposed that entire forests be planted with these trees. One such proposal is for trees which produce no seeds, but divert the energy from seed production to more rapid growth of wood. A forest of such trees would wreak havoc on the food chain. Other GE trees that do produce seeds could cross with native varieties and damage forest ecosystems. Engineered trees which produced pollen (as might happen despite scientists' attempts to create sterile subspecies) could cross with native varieties miles away and damage forest ecosystems.

Fish, as well as other animals, are also being genetically engineered to grow more rapidly. If they are released to the environment (fish culture tanks often discharge during storm conditions), they may out-compete native species and thereby disrupt ecosystems.

There is evidence that soil organisms may be adversely affected bygenetically engineered crops. The Bt corn plant is engineered to contain a bacterial gene that causes production of an insecticide in every cell of the corn plant, including the edible corn ear and the roots. This toxin has been found to persist in the soil for months.

The promoters of genetic engineering show no sign that they are willing or able to impose limits on their applications of this new technology. It will therefore be left to the institutions of civil society—governments working with non-governmental organizations representing the concerned public (such as Sierra Club)—to set limits to how much further genetic engineering will be allowed to alter the earth's species.

Medical Uses

Promoters of the use of genetic engineering outside the laboratory claim that a moratorium or other controls on the planting of genetically engineered organisms as agricultural crops would mean an end to the uses of genetic engineering in the production of medically useful products. This is untrue. As long as proper precautions are taken to assure that the genetically engineered microorganisms used in production of pharmaceuticals or in scientific experiments are not released to the environment, such uses need not be prohibited. However, all applications of genetic engineering, including medical uses, carry some risk. Medical applications of genetic engineering should be approached with caution and not rushed to market. We believe that simpler, more traditional strategies for problem solving should always be considered when evaluating the production of transgenic organisms. This is especially relevant with respect to agricultural applications, as will be discussed in the topic below.

Feeding the World's Hungry?

Medical uses of genetic engineering may be prudent, but agricultural applications of GE are not. Yet the argument is being made by the biotech industry that if the genetic engineering of farm crops is not allowed to proceed, the poor people of the world will starve.

In fact there is more than enough food produced by conventional agriculture, without genetic engineering, to feed all of the world's people. One cause of hunger is the ineffective distribution of food. Genetic engineering may actually lead to more food insecurity and hunger because in poor countries it will lead to the planting of monoculture crops, highly vulnerable to disease and pests, in the place of resilient, diverse range of crops, and it will make farmers dependent on corporations that will demand payment for basic inputs such as seed, chemicals, and fertilizers.

Terms of trade between developed and less developed nations have often resulted in the best land in the poor countries being used to grow cash crops for export rather than food for consumption at home. Issues of equity and fairness have not been addressed by trade agreements. Certainly these problems call out for redress, but their solution isn't to increase the monopoly power of "life science" companies in the richest nations.

As Indian writer and activist Vandana Shiva summarized, "Millions of farmers in third world countries want to breed and grow the crop varieties that adapt to their diverse ecosystems. Plant biodiversity is essential for a balanced diet. Yet numerous crops are pushed to extinction with the introduction of GE crops."

Terminator Technology

Any claim by the corporations promoting agricultural biotechnology that they have the interests of the world's poor people at heart are refuted by the facts in the case of Terminator seed technology. This technology would protect the intellectual property interests of the seed company by making the seeds from a genetically engineered crop plant sterile, unable to germinate. Terminator would make it impossible for farmers to save seed from a crop for planting the next year, and would force them to buy seed from the supplier. In the third world, this inability to save seed could be a major, perhaps fatal, burden on poor farmers. The Sierra Club's Genetic Engineering Committee (GEC) believes that Terminator is a tool by which seed companies are trying to engineer their monopoly power into the genetic code.

Adding insult to injury in the Terminator technology story is the fact that our own tax dollars were used to develop Terminator. The U.S. Dept. of Agriculture played a major role in the development of Terminator technology. The USDA is actually part owner of the patent on this immoral technology.

The Genetic Engineering Committee

The Sierra Club's Genetic Engineering Committee (GEC) was formed in May 1999 to explore ways to mobilize the strength of Sierra Club, the largest grassroots environmental organization in the U.S., for the work of public education and regulatory reform that will be necessary to protect the natural environment and human health from the threats posed by the release of genetically engineered organisms.

An Educational Challenge

Genetic Engineering Committee members have found that the need for public education is great. Because of inadequate reporting by the U.S. media, many otherwise well educated people simply have not been told what genetic engineering is. We hear statements like, "If there is a moratorium on planting genetically engineered crops, doesn't that mean that no crops at all will be planted?" And, "Aren't all farm crops these days genetically engineered?"

The supporters of genetic engineering gladly fill this information vacuum with false statements. They claim that the selective breeding of plants and animals that has been done for centuries is genetic engineering. Supporters claim that modern genetic engineering is nothing more than an improved, more precise, high-tech form of conventional plant and animal breeding. Michael Khoo, in a letter published last year in the Toronto Globe and Mail, called this claim "... biotechnology's public-relations line that genetic engineering is no different from traditional breeding." His letter continued, "A potato can cross with a different strain of potato but, in 10 million years of evolution, it has never crossed with a chicken. Genetic engineering shatters these natural species boundaries, with completely unpredictable results. As a result of these risks, the British Medical Association has recently called for an open-ended moratorium on GE planting." Gene transfers occur in conventional breeding, but these transfers can only take place between individuals of the same species, or, in the case of hybridization, between individuals of closely related species. This is because conventional breeding relies on the normal reproductive processes of the plants or animals. Plants can be conventionally bred only with plants of the same species or, to make a hybrid, with closely related species. And animals can only be bred with other animals of the same or, in some instances, closely related species.

Genetic engineering is not bound by these limits in the possible exchanges of genes that can be made to occur using its techniques, which include the use of viruses as "vectors" to move foreign genes into host organisms. By means of genetic engineering, genes can be transferred from a plant to an animal; from an animal to a plant; from a bacteria to a plant, and between numerous other combinations of donor and recipient organisms. There have even been attempts to put human genes in plants and animals that are used as human food.

Bibliography

Adams, C.E. : *Mammalian Egg Transfer*, Boca Raton, FL, CRC Press, 1982.

Arti Sharma: *Fishes : Aid to Collection, Preservation and Identification*, Daya, Delhi, 2006.

Aruna T. Kumar : *Handbook of Animal Husbandry*, Indian Council of Agricultural Research, 2008.

Baker, Steve: *The Postmodern Animal*. London: Reaktion Books, 2000.

Balram Pani: *Textbook of Animal Chemistry*, I K International, Delhi, 2007.

Basavaraj S. Benni: *Dairy Co-operative Management and Practice*, Rawat, Delhi, 2005.

Bower, B.: *Fossils may Clarify Mammal Evolution*, Science News, 1984.

Brock, J. : *A Natural History of Domesticated Animals*, Cambridge Univ. Pr., New York, 1999.

Bronson, F. H.: *Mammalian Reproductive Biology*, Univ. Chicago Pr., Chicago, 1990.

Clark, Stephen: *The Moral Status of Animals*. Oxford: Oxford University Press, 1977.

Clutton Brock Juliet : *Horse power: a history of the horse and donkey in human societies*, National history Museum publications, London 1992.

Clymer, R. : *Nature's Healing Agents*, PA, U.S.A: Dorrance Co., 1963.

Crawford A. : *Experiments and Observations on Animal Heat*, London: Printed for J. Johnson; 1788.

Daniel, J.C. Jr. : *Methods in Mammalian Reproduction*, Orlando, FL, Academic Press, 1978.

Davis, A. : *Let's Eat Right to Keep Fit*, New York, U.S.A: Harcourt Brace Jovanovich, Inc., 1970.

Degen, A. A. : *Ecophysiology of Small Desert Mammals*, Springer, New York, 1997.

DeGrazia, David: *Animals Rights: A Very Short Introduction*. Oxford: Oxford University Press, 2002.

Devender Pratap Singh : *A Handbook of Beekeeping*, Agrobios, 2006.

Devyani Khemka: *Animal Physiology*, Dominant, Delhi, 2003.

Eisenberg, John E.: *The Animal Radiations*, The University of Chicago Press, 1981.

Ensminger, M.E. : *Dairy Cattle Science*, The Interstate Printers & Publishers, Inc., Danville, 1980.

Escobar, Roberto Calle: *Animal Breeding and Production of Camelids*, Lima, Peru, 1984.

Flowerdew, J. R. : *Animals: Their Reproductive Biology and Population Ecology*, Cambridge Univ. Pr., New York, 1987.

Gay, W.I. : *Methods of Animal Experimentation*, Academic Press, New York, 1965.

Godthelp, *Animal: Riversleigh. The Story of Animals Reproduction in Ancient Rainforests*, Reed Books, Balgowhah, 1991.

Goel, A K : *Basic Concept of Animal Chemistry*, Pearl Books, Delhi, 2008.

Gordon, G. A. : *Animals Physiology*, Harper and Row, New York, 1989.

Gray, J.: *Animal Locomotion*, Norton, New York. 1968.

Hagedorn, A.L.: *Animal Breeding*, Crosby Lockwood, 1950.

Harrison, R. J.: *Functional Anatomy of Marine Animals*, New York: Academic Press, 1974.

Hulbert AJ, Else PL. : *Mechanisms Underlying the Cost of Living in Animals*, Annu Rev Physiol. 2000.

Hutt, Frederick B.: *Genetics for Dog Breeders*, Freeman & Company, 1979.

Lata Bhattacharya: *Animal Biochemistry*, Discovery, Delhi, 2010.

Lyster, S. : *Animals and Their Moral Standing*. London : Routledge, 1997.

Marshall, R.B. : *Breeding Farm Animals*, Asiatic Pub, Delhi, 2006.

Muybridge, E. : *Muybridge's Complete Human and Animal Locomotion*, Dover Publ., New York, 1979.

Nyholt, D.H. : *The Vitamin & Herb Guide*, Alberta, Canada: Global Health Ltd., 1992.

Rathnakumar, K : *Fish Processing Technology and Product Development*, Narendra Pub, Delhi, 2008.

Safley, Michael: *Synthesis of a Miracle*, Northwest Alpacas, 2002.

Schiller, A. L. : *Anatomy of the Guinea Pig*, Harvard Univ. Pr., Cambridge, 1975.

Seidel, S.M.: *New Technologies in Animal Breeding*, Orlando, FL, Academic Press, 1981.

Shukla, M K : *Brain Teasers : Multiple Choice Questions on Animal Husbandry and Veterinary Sciences*, International Book Dist, Delhi, 2007.

Singh,. G : *Chemistry of Amino-Acids and Proteins*, Discovery, Delhi, 2007.

Stuart Patton : *Principles of Dairy Chemistry*, Huntington, N.Y.: Krieger, 1976.

Thornhill, Nancy W.: *The Natural History of Inbreeding and Outbreeding*, Chicago Press, 1993.

Verma, S.R. : *Nature: Fish Genetics and Biodiversity Conservation*, Conservators, Delhi, 1998.

White, M.J.D.: *Animal Cytology and Evolution*, Cambridge, Cambridge Univ. Press, 1954.

Yablokov, A.: *Variability of Mammals: Moscow*, USSR, Nauka Publishers, 1966.

Yadav, Manju: *Mammalian Development*, Discovery Publishing House, Delhi, 2008.

Index